School Science Practical Work in Africa

School Science Practical Work in Africa presents the scope of research and practice of science practical work in African schools. It brings together prominent science educators and researchers from Africa to share their experience and findings on pedagogical innovations and research-informed practices on school science practical work.

The book highlights trends and patterns in the enactment and role of practical work across African countries. Practical work is regarded as intrinsic to science teaching and learning and the form of practical work that is strongly advocated is inquiry-based learning, which signals a definite paradigm shift from the traditional teacher-dominated to a learner-centered approach. The book provides empirical research on approaches to practical work, contextual factors in the enactment of practical work, and professional development in teaching practical work.

This book will be of great interest to academics, researchers and post-graduate students in the fields of science education and educational policy.

Umesh Ramnarain is a Professor in Science Education, and Head of Department in the Department of Science and Technology Education at the University of Johannesburg, South Africa.

Perspectives on Education in Africa

Series Editor: Kerry J Kennedy

The African continent is in a crucial moment of its history. If conflicts, political disappointments, developmental difficulties and poverty issues of Africa are well disseminated by the international media, it should not gloss ever the fact that Africa is also a very dynamic continent, with a promising demography and hopeful economic growth.

Education could be viewed as at the heart of the challenges facing Africa. Schools could offer the promise to achieve the goals of the development, both in social aspects as well as economic and political. Since independences in the 1960's, the number of schoolchildren has multiplied by 40 in sub-Saharan Africa. Many states in Africa, from North to South, are faced with the emergency of mass-schooling while many problems remain: shortage of basic facilities, infrastructure, lack of teaching and learning materials, shortage of qualified teachers, distance between home and schools in rural areas, hunger and poor nutrition, difficulties for schooling in areas affected by conflicts and schooling for girls. Development and improvement in the higher education and vocational training is also a key challenge for African countries, many of which are witnessing the massive student mobility (with its crucial problematics of "brain drain" but also "brain gain"). Some countries stress the need to privatize education to try to achieve international targets. Many of them rely on international support to reach the goals. All these challenges, however, should not obscure the dynamism of African students, the growth of the quality of education in some African countries, such as Morocco, and other visible examples across the Continent.

In focusing on education, the purpose of the proposed Series is to examine an institution that is regarded as fundamental in helping African countries face major challenges across the Continent. *"Education is the most powerful weapon which you can use to change the world"* said Nelson Mandela. This Series will seek to offer tools for analysing, for understanding and for decision-making concerning contemporary issues of Education in Africa.

A basic assumption of the Series is that the perspectives on education in Africa should not be observed, analysed and strategized from outside Africa. The Series will primarily draw on local knowledge and experience within

Africa with the potential to decolonize African education and provide insights by which indigenous knowledge can be promoted and developed. This does not rule out considering perspectives from outside the Continent, especially in the context of globalization but these will not dominate. This Series, however, will also promote interactions between African and non-African scholars in order to explore the implication for education in Africa. Yet the focus will always be on education in and for African people, the way such education can be enhanced, the factors that influence it and future directions in which it can develop.

Books in the series include:

Educational Assessment in a Time of Reform
Standards and Standard Setting for Excellence in Education
Coert Loock and Vanessa Scherman

School Science Practical Work in Africa
Experiences and Challenges
Edited by Umesh Ramnarain

For more information about the series, please visit www.routledge.com/Perspectives-on-Education-in-Africa/book-series/EDUAFRICA

School Science Practical Work in Africa

Experiences and Challenges

Edited by
Umesh Ramnarain

Routledge
Taylor & Francis Group

LONDON AND NEW YORK

First published 2021
by Routledge
2 Park Square, Milton Park, Abingdon, Oxon OX14 4RN

and by Routledge
52 Vanderbilt Avenue, New York, NY 10017

Routledge is an imprint of the Taylor & Francis Group, an informa business

British Library Cataloguing-in-Publication Data
A catalogue record for this book is available from the British Library

Library of Congress Cataloging-in-Publication Data
A catalog record has been requested for this book

ISBN: 978-0-367-20279-8 (hbk)
ISBN: 978-0-367-50538-7 (pbk)
ISBN: 978-0-429-26065-0 (ebk)

Typeset in Bembo
by Taylor & Francis Books

Contents

Illustrations

Contributors

Umesh Ramnarain is a Professor in Science Education, and Head of Department in the Department of Science and Technology Education at the University of Johannesburg, South Africa. His main research interest is in inquiry-based science education, with a particular focus on its uptake in South African classrooms, where the unequal funding policies of the previous Apartheid education system have resulted in learning contexts that are complex and diverse.

Eva Asheela is an Education Officer responsible for curriculum research and development. She works for the Ministry of Education, Arts and Culture at the directorate National Institute for Educational Development (NIED). She is involved in developing the curriculum for science and works with in-service teachers on Continuing Professional Development (CPD).

Kenneth Mlungisi Ngcoza is an Associate Professor in Science Education at Rhodes University. He supervises MEd and PhD students. His research interests include science curriculum, professional development, IKS and ESD. He has published a number of journal articles and is a reviewer for *AJRMSTE, SAJEE, Pythagoras, RSTEJ* and *CSSE*.

Joyce Sewry is a Senior Lecturer in the Department of Chemistry and Deputy Dean of Science at Rhodes University. Her interests are in community engagement and maths and science education. She has been awarded the Vice Chancellor's Distinguished Award for both community engagement (which she shared with Ken Ngcoza) and teaching.

Lydia Mavuru is a Senior Lecturer in Science Education at the University of Johannesburg, South Africa. Her main research interest is socio-cultural perspectives in science education, focusing on social constructivist pedagogies and consideration of equity and diversity in making science comprehensible and relevant to learners.

Washington T. Dudu is the Deputy Dean of Research and Innovation in the Faculty of Education, North-West University. Washington's research interests are in scientific inquiry, nature of science, pedagogical content knowledge and the affordances of indigenous knowledge in the science classroom.

Johnson Enero Upahi is a postdoctoral research fellow in the Department of Science and Technology Education, University of Johannesburg, South Africa. His research interests focuse on context and problem-based learning, nature of science and scientific literacy, science curricula and textbook analysis, inquiry-based science education, and chemistry teachers' pedagogical content knowledge.

Oloyede Solomon Oyelekan obtained his PhD and lectures in the Department of Science Education, University of Ilorin, Nigeria. In 2009, his PhD thesis won a national award from the National Universities Commission as the best in the field of education. One of his current research interests is on how to use ICT and virtual reality to improve students' learning of science.

Dorothy C. Nampota is a Professor of Science Education at Chancellor College of the University of Malawi. In the various education sectors she has worked with, she has achieved a lot, such as enhancing girls' participation and performance in science and mathematics. Her research interests focus on understanding the teaching and learning process in science and developing interventions for effective science teaching.

Nellie M. Mbano is a Senior Lecturer in Biology Education at Chancellor College. Her interests include teaching thinking, investigative science, using active learning, and gender in science education

Bob Maseko is a Lecturer in Science Education at the University of Malawi Chancellor College. Currently, he is a PhD student at the University of Wits in South Africa. He received his BEd and MEd in Science Education from the University of Malawi and Leeds, respectively. His research interests include the development and enactment of PCK in different classroom contexts.

Josephat M. Miheso is a seasoned curriculum developer, currently the Head of Department Mathematics and Science and Chairperson of the Regional Center of Excellence (RCoE) on Curriculum Matters at the Kenya Institute of Curriculum Development (KICD), Nairobi, Kenya. His research interests are in professional teacher knowledge for teaching science, with both practising and preservice teachers.

Israel Kibirige is a Research Associate Professor at the University of Limpopo. He has published 30 articles in peer-reviewed journals. He was the Research Chair in "Quality Teaching and Learning" at the University of Limpopo, 1912–2017. His research interests are in estuarine ecology, stable isotopes, and science education.

Vivien Mweene Chabalengula is Associate Professor of Science Education at the University of Virginia. Prior, she taught high school biology and chemistry in Zambia, and science education courses at the University of Zambia. Her primary research focus is on promoting scientific literacy among students, and analysing STEM curriculum materials for various aspects including, but not limited to, the integration of scientific literacy themes, science and engineering design practices and design skills, real-world applications, and multicultural science contexts.

Frackson Mumba is Associate Professor of Science Education at the University of Virginia, USA. His research is on inquiry instruction, engineering design integration in science teaching and its impact on student learning, and pedagogical content knowledge for chemistry teaching at college level. His research work has resulted in several presentations at international and national conferences, and publications in peer-reviewed science and engineering education journals.

Ladislaus M. Semali is Professor of Education (Curriculum and Instruction; Adult Education, & Comparative and International Education). Dr. Semali is concurrently Professor of Education at Mwenge Catholic University and Professor Emeritus of Education at the Penn State University. Dr. Semali has extensive experience in organizing internships and workshops for adult learners in Africa and is currently directing research on the existentialist crisis of interest in STEM education in East Africa.

Series Editor's foreword

The African continent is in a crucial moment of its history. If conflicts, political disappointments, developmental difficulties and poverty issues of Africa are well disseminated by the international media, it should not gloss over the fact that Africa is also a very dynamic continent, with a promising demography and hopeful economic growth.

Education could be viewed as at the heart of the challenges facing Africa. Schools could offer the promise to achieve the goals of development, both in social aspects as well as economic and political. Since the independences of the 1960s, the number of schoolchildren has multiplied by 40 in sub-Saharan Africa. Many states in Africa, from North to South, are faced with the emergency of mass schooling while many problems remain: shortage of basic facilities, infra-structure, lack of teaching and learning materials, shortage of qualified teachers, distance between home and schools in rural areas, hunger and poor nutrition, difficulties for schooling in areas affected by conflicts and schooling for girls. Development and improvement in higher education and vocational training is also a key challenge for African countries, many of which are witnessing massive student mobility (with its crucial problematics of "brain drain" but also "brain gain"). Some countries stress the need to privatize education to try to achieve international targets. Many of them rely on international support to reach the goals. All these challenges, however, should not obscure the dynamism of African students and the growth of the quality of education in some African countries, such as Morocco, and other visible examples across the continent.

In focusing on education, the purpose of the proposed series is to examine an institution that is regarded as fundamental in helping African countries face major challenges across the continent. As Nelson Mandela said, "Education is the most powerful weapon which you can use to change the world". This series will seek to offer tools for analysing, understanding and decision-making with regard to the contemporary issues of education in Africa.

A basic assumption of the series is that the perspectives on education in Africa should not be observed, analysed and strategized from outside Africa. The series will primarily draw on local knowledge and experience within Africa with the potential to decolonize African education and provide insights through which indigenous knowledge can be promoted and developed. This does not rule out considering perspectives from outside the continent, especially in the context of globalization, but these will not dominate. This series, however, will also promote interactions between African and non-African scholars in order to explore the implications for education in Africa. Yet the focus will always be on education in and for African people, the way such education can be enhanced, the factors that influence it and future directions in which it can develop.

The present book on engagement in practical science activities deals with issues that are fundamental, not just to education but to economic development as well. The pan-African sweep of the book provides insights that can help position African countries for the future. It is a welcome addition to the international literature.

Kerry J Kennedy
Series Editor
Joseph Divala, Juliet Perumal, & Elizabeth Henning
Co-Series Editors

Chapter 1

Inquiry-based learning in South African schools

Umesh Ramnarain

Introduction

One of the key imperatives in the transformation of education in South Africa is the need to provide quality education for all (Department of Education, 2001). A framework for transformation of the education system is the Department of National Education's *White Paper 1 on Education and Training* (1994), which articulates the main objective for science education: the improvement in the quality of school science for Black students. Research on the impact of inquiry-based learning suggests that this approach can result in an improvement in science performance. A number of studies provide evidence that teaching science using a scientific inquiry method can improve learners' performance in science (Maxwell & Lambert, 2015; Minner, Levy & Century, 2010). A strong force giving impetus to change in science education was the assertion that the previous curriculum was both inaccessible and irrelevant to Black students (Naidoo & Lewin, 1998). One of the major changes advocated in this curriculum reform is a new conception of the role and form that practical work should assume. The importance that is given to practical work is highlighted in the new Curriculum and Assessment Policy Statement (CAPS) where it is stated that practical work "must be integrated with theory to strengthen the concepts being taught" (Department of Basic Education, 2011, p. 11).

Inquiry-based science education is posited as the means by which the challenges of the previous curriculum related to inaccessibility, irrelevance and incompatibility with the nature of science can be negotiated (Department of Basic Education, 2011). Scientific inquiry has been advocated as a common curriculum goal in school science education in South Africa, and also throughout the world. Inquiry-based learning allows learners to develop "key scientific ideas through learning how to investigate and build their knowledge and understanding of the world" by using "skills employed by scientists such as raising questions, collecting data, reasoning and reviewing evidence in the light of what is already known, drawing conclusions and discussing results" (Inter-Academy Panel, 2012, p. 19). Inquiry is a multifaceted activity, the essence of which is captured in the following widely quoted description in the National Science Education Standards of the United States:

> Inquiry is a multifaceted activity that involves making observations; posing questions; examining books and other sources of information to see what is already known; planning investigations; reviewing what is already known in light of experimental evidence; using tools to gather, analyze, and interpret data; proposing answers, explanations, and predictions; and communicating the results. Inquiry requires identification of assumptions, use of critical and logical thinking, and consideration of alternative explanations.
>
> (NRC, 1996, p. 23)

In South Africa, the National Curriculum and Assessment Policy Statement (CAPS) is a single, comprehensive and concise policy document introduced by the Department of Basic Education that gives detailed guidance for teachers on what they should teach and how to assess. One of the principles of the curriculum is "Active and critical learning: encouraging an active and critical approach to learning, rather than rote and uncritical learning of given truth" (Department of Basic Education, 2011, p. 4). Inquiry-based learning is an approach that gives expression to this principle in science classroom teaching and learning practice. In South Africa, this notion of an inquiry-based science curriculum is underlined through the statement of curricular aims. Specific Aim One states that "the purpose of Physical Sciences is to make learners aware of their environment and to equip learners with investigating skills relating to physical and chemical phenomena" (Department of Basic Education, 2011, p. 8). This curriculum goal is also highlighted in Specific Aim Two of the same CAPS document where it is stated that Physical Sciences "promotes knowledge and skills in scientific inquiry and problem solving; the construction and application of scientific and technological knowledge; an understanding of the nature of science and its relationships to technology, society and the environment" (Department of Basic Education, 2011, p. 8). Similar aims that reinforce the idea of an inquiry-based pedagogy are reflected in curriculum documents for other science subjects such as Life Sciences and Natural Sciences.

The curricular underpinnings of an inquiry-based approach become most evident when this approach is contrasted with a traditional approach to science teaching. The traditional science curriculum, which placed much emphasis on the transmission of scientific knowledge, was teacher-centred, and portrayed the learner in a passive role. Here, learners slavishly follow teacher directions and procedures without much thought (Hodson, 1993). Experimental tasks in this mode often embody a cookbook approach, where learners followed recipes for the execution of procedures handed down by teachers, and gathered and recorded data without a clear sense of purpose (Roth, 1994). In such a teacher-centred science classroom, communication flows from the teacher to the learner and teacher talk dominates the lesson. It was anticipated that the infusion of inquiry-based learning would redefine this prevailing science teacher–learner relationship and, thereby, shift the communication pattern in the classroom towards more learner-centredness.

In a typical South African classroom, learners might sit in straight rows of desks facing the front of the class and have few opportunities to interact or work in collaborative learning groups. Many of the activities carried out by learners merely confirm or illustrate science concepts, laws or principles (Hobden, 2005). Although these prescriptive exercises teach basic science process skills such as observing, inferring, measuring, communicating, classifying and predicting, the most crucial drawback of such an approach is that it does not address the conceptual, epistemic, social and/or procedural domains of scientific knowledge. A pedagogical framework on inquiry developed by van Uum, Verhoeff and Peeters (2016) describes how these four domains may be accessed in the different phases of inquiry learning. The conceptual domain of science describes a "body of knowledge that represents current understanding of natural systems" (NRC, 2007, p. 26). The epistemic domain refers to the nature of science and the way scientific knowledge is generated (Duschl, 2008). The social domain of science refers to research collaboration and communication, and the critical review of work within a disciplinary community of practice (Furtak et al., 2012). The procedural domain addresses inquiry procedures, such as formulating research questions and drawing conclusions to answer the research questions (Furtak et al., 2012). The traditional practice of practical work in South Africa that is characterized by the 'cookbook' approach has denied learners access to the tenets of science encapsulated within these four domains of knowledge. This conception of practical work was not compatible with the nature of science. The learners were exposed only to the products of the scientific enterprise in the form of facts, concepts, principles and laws of the physical world. This knowledge is referred to as the substantive aspects of science. This static view of science has in no small part contributed to the rote learning in South African science classrooms.

Scientific inquiry can provide a viable context for addressing the nature of science in the classroom (Schwartz & Crawford, 2006), amongst other benefits. Studies worldwide have reported the benefits of inquiry-based teaching and learning. These benefits include stimulating an interest in science and increased motivation (Potvin & Hasni, 2014), improved understanding of concepts (White & Frederiksen, 1998), an understanding of the nature of science (Gaigher, Lederman & Lederman, 2014; Schwartz & Crawford, 2006), the development of higher-order thinking (Conklin, 2012), and facilitating collaboration between learners (Hofstein & Lunetta, 2003). Accordingly, these benefits have been recognized by science teachers in South Africa. A study by Ramnarain (2010b) reported on how teachers and learners perceive the benefits of autonomous science investigative inquiries in the grade 9 (age 13–14 years) in Natural Sciences. The study adopted a mixed methods research design involving the collection of both quantitative and qualitative data. The three perceived benefits reported are that it is motivational, it facilitates conceptual understanding, and it leads to the development of scientific skills.

Another study investigated the effect of inquiry-based learning on the achievement goal orientation of grade 10 Physical Sciences learners at historically disadvantaged township schools in South Africa (Mupira & Ramnarain, 2018). Achievement goal theory focuses on understanding the different goals in learning and identifies two main goal orientations. The first is mastery goal orientation where the intrinsic value of learning is key (Meece, Herman & McCombs, 2003) and the focus is on the challenge and mastery of a science task (Velayutham, Aldridge & Fraser, 2012). Students with a mastery goals orientation are not concerned about how many mistakes they make or how they appear to others but view mistakes as learning opportunities and as something that can help them to learn (Koballa & Glynn, 2007). In contrast to a mastery goal orientation, students who adopt performance goals are expected to minimally persist in the face of difficulty, avoid challenging tasks, and to have low intrinsic motivation (Ames, 1992). The findings showed that learners who experienced inquiry-based learning significantly gained in mastery goal orientation, while the control group that were taught through a traditional direct didactic approach showed insignificant change in their mastery goal orientation. From these results, it can be concluded that inquiry-based learning does support a mastery goal orientation in learners. This orientation is regarded as desirable because mastery approach goals could support positive outcomes in conceptual learning, leading to an improvement in the science achievement of learners.

Emerging framework for types of inquiry in South Africa

Various models and classification frameworks have been presented for inquiry-based learning that can help teachers organize and sequence investigative learning experiences for their learners. By means of a model, the features of scientific inquiry may be combined in a series of coherent learning experiences that help learners build new understandings and develop their investigative skills over time (NRC, 2000). They also provide opportunities for learners to extend, apply and evaluate what they have learned (Bybee, 1997). A key feature in each of these models and frameworks is the openness or closure of the inquiry. In terms of the degree of learner autonomy and the extent of teacher control, scientific inquiry may lie along a spectrum from open to closed depending upon who makes the decisions in the investigation process (Abraham, 1982; Hackling & Fairbrother, 1996).

Informed by the preceding literature and guided by the outcomes and assessment standards of the new curriculum, the author formulated a classification framework for inquiry types (Ramnarain & Hobden, 2015). In this framework, five stages of inquiry are reflected, namely: choosing the topic; formulating the question; planning; collecting data; and analysing data and drawing a conclusion. The five stages indicated in the framework cohere well with scientific inquiry skills such as "identifying problems and issues, raising

Table 1.1 Framework for types of inquiry

Inquiry level	Inquiry stages				
	Choosing the topic	Formulating the question	Planning data collection	Collecting data	Analysing data and drawing a conclusion
1	T	T	T	T	L
2	T	T	T	L	L
3	T	T	L	L	L
4	T	L	L	L	L
5	L	L	L	L	L

L = Learner has autonomy and responsibility to carry out the inquiry
T = Teacher controls and carries out the inquiry

questions, planning, doing investigations, interpreting data, and communicating results and conclusions" that are specified for assessment in the South African Curriculum and Assessment and Policy Statement (CAPS) for Physical Sciences (Department of Basic Education, 2011, p. 16). Based on the degree of autonomy that is entrusted to learners, and the extent of teacher involvement, the inquiry can be classified into five levels. These levels are described in terms of the interplay between learner autonomy and teacher control over the stages of inquiry.

For example, an inquiry at level 2 means that the teacher chooses the topic, provides a question to investigate and provides a plan for data collection, while the learner is entrusted with collecting the data, and analysing the data and drawing a conclusion. The classification framework constructed is useful as it allows for the type of inquiry to be determined based on the degree of autonomy given to learners at each of the stages. It was felt that such differentiation was necessary so that the openness or closure of the inquiry could be described in a clear and unambiguous manner.

Implementation of inquiry-based learning and a proposed learning progression

Research that has been conducted by the author suggests that learners enjoyed only limited autonomy in the enactment of inquiry. For example, in a survey of 55 teachers of grade 9 Natural Sciences, it was found that scientific inquiries are largely at levels 1 and 2, where the teacher exercises a great deal of control over the stages of the investigation (Ramnarain, 2010a). This trend is corroborated in a study by Dudu (2017) on a surveyed sample of grade 11 Physical Sciences learners where it was found that they experienced generally moderate levels of inquiry. However, what also emerges from research is that there is no uniformity in the extent and type of inquiry implemented across South African schools. This is evident in a study that investigated the perceptions of Physical

Sciences teachers regarding the implementation of inquiry-based learning in a diverse range of high schools in South Africa (Ramnarain, 2014b). The findings show that teachers at all school locations had a positive perception of inquiry-based learning, with perceived benefits for learners including the development of experimental skills and making science more enjoyable. However, with regard to inquiry facilitating conceptual understanding, teachers at township and rural schools believed a didactic approach would be more effective than learners doing inquiry, while teachers at suburban and urban schools favoured an inquiry-based approach in this regard. The significance of this study is that the lack of resources, large classes, and learners' limited exposure to inquiry in township and rural schools constrain the implementation of inquiry-based learning at these schools, and result in teachers at such schools resorting to a didactic pedagogy.

The findings of this study on the role played by context in the enactment of inquiry were corroborated in another study that investigated the pedagogical orientations of in-service Physical Sciences teachers at a diverse range of schools in South Africa (Ramnarain & Schuster, 2014). The findings reveal remarkable differences between the orientations of teachers at disadvantaged township schools compared to those at more privileged suburban schools. It was found that teachers at township schools have a strong 'active direct' teaching orientation overall, involving direct exposition of the science followed by confirmatory practical work, while teachers at suburban schools exhibit a guided inquiry orientation, with concepts being developed via a guided exploration phase. The study again identified contextual factors such as class size, availability of resources, time constraints, student ability, school culture and parents' expectations as influencing the methods adopted by teachers. It would also appear that the influence of these factors is most acute in rural schools. This was revealed in a study by Ramnarain and Hlatswayo (2018) that investigated the interactions between the beliefs of grade 10 rural Physical Sciences teachers about inquiry-based learning, and their practice of inquiry in their classrooms. The findings reflect that the sampled teachers from the rural district have a positive attitude towards inquiry in the teaching and learning of Physical Sciences, and recognize the benefits of inquiry, such as addressing learner motivation and supporting learners in the understanding of abstract science concepts. However, despite this positive attitude towards inquiry-based learning, teachers are less inclined to enact inquiry-based learning in their lessons due to the aforementioned factors.

Aside from these extrinsic factors, a further study revealed that a lack of professional science knowledge (content knowledge, pedagogical content knowledge, pedagogical knowledge, knowledge of students, educational contexts, curricular knowledge, and educational purposes) contributes towards teachers' uncertainty in inquiry-based teaching (Ramnarain, 2016).

What is clear from all these studies is that a stepped approach is needed in steering learning towards guided and open inquiries. In this approach, the role of the teacher in scaffolding learner progress in the inquiry is key. Gabel (2001) describes scaffolding as "a bridge used to build upon what students already know to arrive at something that they do not know" (p. 61). Stemming from the research by Ramnarain and Hobden (2015) on scaffolding strategies used by teachers of grade 9 Natural Sciences at the various stages of inquiry is the notion of a learning progression of inquiry-based practice, which can offer planned support in facilitating learners' progress towards greater autonomy. The learning progression is based on an inverse relationship between teacher support and learner autonomy. Teacher support at the various stages of the inquiry can be in the form of skills development, a structured prompt sheet, suggestions and hints, and probing questions. The learning progression proposes that when learners function at autonomy level 1, where they may have little or no experience of inquiry, optimal scaffolding is need to develop capacity. Here, teachers may support learners by giving them exercises in formulating questions, identifying variables, using apparatuses, and drawing graphs. As learners develop competency and confidence in doing inquiry, the teacher support is gradually withdrawn, such that when they reach level 5 of inquiry they are sufficiently empowered to conduct an open inquiry. This learning progression is a guide by which teachers can gradually shift their inquiry-based practice towards one characterized by greater learner autonomy.

Assessing inquiry

Assessment is a topic that is foremost in the discourse of South African teachers, given the emphasis on summative assessment. This is most strongly evident in the attention given to preparing learners for high stakes assessment such as the exit examination that is taken by grade 12 learners. The assessment of inquiry learning is therefore a key consideration within the South African context. The CAPS document refers to a programme of continuous formal assessment for grade 12 which makes up 25% (100 marks) of the total mark for Physical Sciences, with 75% (300 marks) being allocated to the final external examination. The 100 marks for continuous assessment is comprised of 45 marks dedicated to experiments. For grade 12, as a part of formal assessment, three experiments are prescribed per year, while for grades 10 and 11, two prescribed experiments are conducted per year, with one being a Physics experiment and the other being a Chemistry experiment. However, there is little specification on how these practical tasks should be assessed.

The issue of assessing inquiry learning has stimulated much debate amongst scholars in science education. It has been contended that traditional forms of assessment in the form of paper-and-pencil items that feature in summative assessments such as tests and examinations fail to capture the complexity of inquiry-based learning (Buckley et al., 2010). However, Ketelhut et al. (2005)

express the view that if teachers are to reinforce an inquiry-based pedagogy in their classrooms, then there needs to be more emphasis on inquiry-based assessment in standardized testing such as examinations.

A key issue in the quality of inquiry assessments is validity. According to Messick (1989), validity is an "evaluative judgement of the degree to which empirical evidence and theoretical rationales support the adequacy and appropriateness of inferences and actions based on test scores or other modes of assessment" (p. 13). In essence, this means the extent to which an assessment task accurately measures what it is supposed to measure.

A study by Ramnarain (2014a) investigated the validity of inquiry tasks in national grade 12 Physical Sciences examinations. This study revealed that inquiry questions in the national Physical Sciences examinations lack construct validity. In certain cases it was not clear what construct was being targeted in the question, and as a result no valid inferences could be made on learner performance. There were also items where the supposed construct that was being targeted was not really being addressed, and hence displayed construct irrelevance. The findings of this study suggest that greater attention needs to be given to the formulation of inquiry-related questions in written tests and examinations.

Teacher professional development in inquiry-based teaching

As mentioned already, a key factor in the effective implementation of inquiry-based science education in South Africa is that teachers have a knowledge deficit in inquiry-based science education. Teachers' understanding of inquiry is regarded as being critical for the enactment of an inquiry-based pedagogy (Abd-El-Khalick, 2013; Ozel & Luft, 2013). Teachers have difficulty adopting an inquiry-based approach to practical work that is in stark contrast to the traditional 'cookbook' approach where students follow 'recipes' for the execution of procedures handed down by the teacher without much thought (Kim & Tan, 2010; Millar, 2010). According to Crawford (2012), "inquiry-based teaching is a complex and sophisticated way of teaching that demands significant professional development" (p. 292). Huge investment has been put into professional development programmes offered by the South African Department of Basic Education to support teachers in inquiry-based teaching; however, there is paucity of evidence to suggest it has gained traction in South African classrooms (Dudu & Vhurumuku, 2012; Ramnarain, 2016). Two reasons can be offered for the lack of effectiveness of the professional support offered. One is that the support has been in the form of a one-shot workshop where teachers are instructed on what to do without being engaged to critically reflect on what form and level of inquiry may be suitable for the context in which they are teaching. As a result, the support has been decontextualized from the teaching scenario in which teachers find themselves. A second criticism of this support has been that is not sustainable. There is very little or no follow-up with teachers on the extent to which they are able to adopt inquiry and the challenges they encounter in its uptake.

Research by the author and fellow researchers has explored an empowerment evaluation approach in capturing, portraying and developing the pedagogical practice of science teachers in inquiry-based teaching (Ramnarain & Makhubalo, 2018; Ramnarain & Modiba, 2013). Empowerment evaluation is posited as an approach that can be exploited in South African schools to facilitate an on-the-job, self-initiated professional development that is environmentally conscious and ongoing. It is an approach whereby individuals can achieve self-determination in their practice (Fetterman, 1999). Fetterman (1999) describes it as a "stepped approach" (p. 16). Empowerment evaluation commences with 'taking stock' of the teacher's current practice (Fetterman, 2001). In 'taking stock', a baseline is established to measure future progress. This is followed by setting realistic and immediate goals, based on consensual agreement between the teacher and the researcher. The third step involves participating teachers selecting and developing strategies to accomplish the set goals. This is achieved through the process of brainstorming, critical review and consensual agreement (Fetterman, 1999). The final step is to 'document progress' that has been made in reaching the goals that were set. The principles of empowerment evaluation are in sync with the characteristics of effective professional development described by Darling-Hammond and McLaughlin (1995) and Loucks–Horsley et al. (1998). For example, Darling-Hammond and McLaughlin (1995) state that professional development should engage teachers in concrete tasks of teaching, assessment, observation, and reflection. This characteristic is a key feature of empowerment evaluation. Loucks–Horsley et al. (1998) maintain that effective professional development builds learning communities where continued learning is valued, and again this feature is evident in empowerment evaluation through its emphasis on sustainable development.

Research conducted in South Africa has shown that this approach has been effective in shifting the practice of teachers towards an inquiry-based pedagogy. In a case study conducted by the author and a fellow researcher (Ramnarain & Modiba, 2013), an empowerment evaluation approach was used to enable a Life Sciences teacher to reflect upon and refine his curriculum design principles in promoting the scientific literacy of learners through inquiry-based learning. By means of stimulated recall discussions that were facilitated by the researchers, the teacher was able to examine the merit of his current practice, and then make a conscious decision to shift his practice from a teacher-centred pedagogy to a learner-centred inquiry-based approach. In other research, a mixed methods multiple case study design was used in investigating, in an empowerment evaluation approach, the professional development of three Physical Sciences teachers towards an inquiry-based pedagogy (Ramnarain & Makhubalo, 2018). Data were collected by means of semi-structured interviews, the Pedagogy of Science Teaching Test-Physical Sciences (POSTT-PS) instrument, and classroom observations by applying the Electronic Quality of Inquiry Protocol (EQUIP) as a classroom observation tool. The findings reveal how over a series of eight lessons through empowerment evaluation, teacher practices in the classroom shifted towards an inquiry-based approach. This shift was reflected in changes to pedagogical orientation, instructional approach, classroom discourse, and assessment practices.

The two studies reported here have shown the potential of this approach in the South African educational landscape in supporting teachers to adopt a stepped approach towards an innovative inquiry-based pedagogy.

Conclusion

This chapter has presented an overview of inquiry-based school science practical work within the South African landscape. The chapter draws upon research that recounts the enactment of inquiry-based teaching and learning in schools. What becomes evident from this review of research is that the unequal funding policies of the previous Apartheid education system have resulted in a learning and teaching terrain that is complex and diverse, giving rise to inherent and inextricable context-dependent factors as they pertain to inquiry-based science education. The research that is reported shows that the implementation, teacher perceptions and pedagogical orientations towards inquiry is a function of both intrinsic teacher factors such as competency and efficacy, and extrinsic school factors such as resources and class size. For the majority of South African teachers, the reformed curriculum has been a paradigmatic shock: a serious departure from deeply entrenched practices characterized by teacher-centredness and learner passivity. Further to this, teachers have a reluctance to 'open up' the learning space for inquiry due to a lack of science subject matter, understanding of the nature of scientific inquiry, and pedagogical knowledge on how to create an inquiry-based learning environment in the classroom. Professional development efforts in South Africa do not pass muster because they have failed to embody features such as sustainability, self-determination, collaboration, contextualization and teacher reflection. All these aspects are reflected in the principles of empowerment evaluation that have been adopted by the author and his fellow researchers in research on science teacher development in South Africa. The learning progression that is proposed can also be adopted as a guideline for teachers on how they may gradually relinquish their control of the learning environment and how learners may be scaffolded to higher levels of autonomy when doing inquiry.

References

Abd-El-Khalick, F. (2013). Teaching with and about nature of science, and science teacher knowledge domains. *Science and Education*, 22(9), 2087–2107.

Abraham, M. J. (1982). A descriptive instrument for use in investigating science laboratories. *Journal of Research in Science Teaching*, 19(2), 155–165.

Ames, C. (1992). Classrooms: Goals, structures, and student motivation. *Journal of Educational Psychology*, 84, 261–271.

Buckley, B. C., Gobert, J., Horwitz, P., & O'Dwyer, L. (2010). Looking inside the black box: Assessing model-based learning and inquiry in BioLogica. *International Journal of Learning Technologies*, 5(2), 166–190.

Bybee, R. W. (1997). *Achieving scientific literacy: From purposes to practices*. Portsmouth: N. H. Heinemann.

Conklin, W. (2012). *Higher order thinking skills to develop 21st century learners*. Huntington Beach, CA: Shell Education.

Crawford, B. A. (2012). Moving the essence of inquiry into the classroom: Engaging teachers and students in authentic science. In K. Tan & M. Kim (Eds.), *Issues and challenges in science education research* (pp. 25–42). Dordrecht: Springer.

Darling-Hammond, L., & McLaughlin, M. W. (1995). Policies that support professional development in an era of reform. *Phi Delta Kappan*, 76(8), 597–604.

Department of Basic Education. (2011). *Curriculum and assessment policy statement: Grades 10–12 physical sciences*. Pretoria: Government Printer.

Department of Education. (2001). *Education in South Africa: Achievements since 1994*. Pretoria: Government Printer.

Department of National Education. (1994). *White paper 1 on education and training*. Pretoria: Government Printer.

Dudu, W., & Vhurumuku, E. (2012). Teachers' practices of inquiry when teaching investigations: A case study. *Journal of Science Teacher Education*, 23, 579–600.

Dudu, W. (2017). Facilitating small-scale implementation of inquiry-based teaching: Encounters and experiences of experimento multipliers in one South African province. *International Journal of Science and Mathematics Education*, 15(4), 625–634.

Duschl, R. A. (2008). Science education in three-part harmony: Balancing conceptual, epistemic, and social learning goals. *Review Research in Education*, 32(1), 268–291.

Fetterman, D. (1999). Reflections on empowerment evaluation: Learning from experience. *Canadian Journal of Program Evaluation, Special Issue*, 5–37.

Fetterman, D. (2001). *Foundations of empowerment evaluation*. Thousand Oaks, CA: Sage.

Furtak, E. M., Seidel, T., Iverson, H., & Briggs, D. C. (2012). Experimental and quasi-experimental studies of inquiry-based science teaching: A meta-analysis. *Review of Educational Research*, 82(3), 300–329.

Gabel, C. (2001). *Effectiveness of a scaffolded approach for teaching learners to design scientific inquiries*. Unpublished doctoral dissertation, University of Colorado, Denver.

Gaigher, E., Lederman, N., & Lederman, J. (2014). Knowledge about inquiry: A study in South African high schools. *International Journal of Science Education*, 36(18), 3125–3147.

Hackling, M. W., & Fairbrother, R. W. (1996). Helping students to do open investigations in science. *Australian Science Teachers' Journal*, 42(4), 26–33.

Hobden, P. A. (2005). What did you do in science today? Two case studies of grade 12 physical science classrooms. *South African Journal of Science*, 101, 302–308.

Hodson, D. (1993). Re-thinking old ways: Towards a more critical approach to practical work in school science. *Studies in Science Education*, 22(2), 85–142.

Hofstein, A., & Lunetta, V. (2003). The laboratory in science education: foundations for the twenty-first century. *Science & Education*, 88, 28–53.

Inter-Academy Panel. (2012). *Taking inquiry-based science education into secondary education*. Report of a global conference. Available from www.fondation-lamap.org/fr/printpdf/16869

Ketelhut, D. J., Clarke, J., Dede, C., Nelson, B., & Bowman, C. (2005, April 4–8). *Inquiry teaching for depth and coverage via multi-user virtual environments.* Paper presented at the National Association for Research in Science Teaching, Dallas.

Kim, M., & Tan, A. L. (2011). Rethinking difficulties of teaching inquiry-based practical work: Stories from elementary pre-service teachers. *International Journal of Science Education,* 33, 465–486.

Koballa, T. R. J., & Glynn, S. M. (2007). Attitudinal and motivational constructs in science learning. In S. K. Abell & N. G. Lederman (Eds.), *Handbook of research on science education* (pp. 75–102). New York: Routledge.

Loucks-Horsley, S., Hewson, P. W., Love, N., & Stiles, K. E. (1998). *Designing professional development for teachers of mathematics and science.* Thousand Oaks, CA: Corwin Press.

Maxwell, D. O., & Lambert, D. T. (2015). Effects of using inquiry-based learning on science achievement for fifth-grade students. *Asia-Pacific Forum on Science Learning and Teaching,* 16(1), 1–29.

Meece, J. L., Herman, P., & McCombs, B. (2003). Relations of learner-centered teaching practices to adolescents' achievement goals. *International Journal of Educational Research,* 39, 457–475.

Messick, S. (1989). Validity. In R. Linn (Ed.), *Educational measurement* (3rd ed., pp. 13–104). Washington, DC: American Council on Education and Macmillan.

Millar, R. (2010). *Analysing practical science activities to assess and improve their effectiveness.* (Getting practical). Hatfield: Association for Science Education.

Minner, D. D., Levy, A. J., & Century, J. (2010). Inquiry-based science instruction – what it is and does it matter? Result from a research synthesis years 1984–2002. *Journal of Research in Science Teaching,* 47(4), 474–496.

Mupira, P., & Ramnarain, U. (2018).The effect of inquiry-based learning on the achievement goal-orientation of grade 10 physical sciences learners at township schools in South Africa. *Journal of Research in Science Teaching,* 55(6), 810–825.

Naidoo, P., & Lewin, J. (1998). Policy and planning of physical science education in South Africa: Myths and realities. *Journal of Research in Science Teaching,* 35(7), 729–744.

National Research Council (NRC). (1996). *National science education standards.* Washington, DC: National Academy Press.

National Research Council (NRC). (2000). *Inquiry and the national science education standards: A guide for teaching and learning.* Washington, DC: National Academy Press.

National Research Council (NRC). (2007). *Taking science to school: Learning and teaching science in grades K–8.* Washington, DC: National Academies Press.

Ozel, M., & Luft, J. A. (2013). Beginning secondary science teachers' conceptualization and enactment of inquiry-based instruction. *School Science and Mathematics,* 113(6), 308–316.

Potvin, P., & Hasni, A. (2014). Interest, motivation and attitude towards science and technology at K-12 levels: A systematic review of 12 years of educational research. *Studies in Science Education,* 50(1), 85–129.

Ramnarain, U. (2010a). A report card on learner autonomy in science investigations. *African Journal of Research in Mathematics, Science and Technology Education,* 14(1), 61–72.

Ramnarain, U. (2010b). Grade 9 science teachers' and learners' appreciation of the benefits of autonomous science investigations. *Education as Change,* 14(2), 187–200.

Ramnarain, U. (2012). The readability of a high stakes physics examination. *Acta Academica,* 44(2), 110–129.

Ramnarain, U. (2014a). Questioning the validity of inquiry assessment in a high stakes physical sciences examination. *Perspectives in Education*, 32(1), 179–191.

Ramnarain, U. (2014b). Teachers' perceptions of inquiry-based learning in urban, suburban, township, and rural high schools: The context-specificity of science curriculum implementation in South Africa. *Teaching and Teacher Education*, 38, 65–75.

Ramnarain, U., & Hlatswayo, M. (2018). Teacher beliefs and attitudes about inquiry-based learning in a rural school district in South Africa. *South African Journal of Education*, 38(1), 1–10.

Ramnarain, U., & Hobden, P. (2015). Shifting South African learners towards greater autonomy in scientific investigations. *Journal of Curriculum Studies*, 47(1), 94–121.

Ramnarain, U. (2016). Understanding the influence of intrinsic and extrinsic factors on inquiry-based science education at township schools in South Africa. *Journal of Research in Science Teaching*, 53(4), 598–619.

Ramnarain, U., & Makhubalo, N. (2018). *An empowerment evaluation approach in shifting a South African science teacher towards an inquiry-based pedagogy*. A paper presented at the International Science Education Conference, 19–21 June, National Institute of Education, Singapore.

Ramnarain, U., & Modiba, M. (2013). Critical friendship, collaboration and trust as a basis for self-initiated professional development: A case of science teaching. *International Journal of Science Education*, 35(1), 65–85.

Ramnarain, U., & Schuster, D. (2014). The pedagogical orientations of South African physical sciences teachers toward inquiry or direct instructional approaches. *Research in Science Education*, 44(4), 627–650.

Roth, W-M. (1994). Experimenting in a constructivist high school laboratory. *Journal of Research in Science Teaching*, 31(2), 197–223.

Schwartz, R. S., & Crawford, B. A. (2006) Authentic scientific inquiry as context for teaching nature of science: Identifying critical element. In L. B. Flick & N. G. Lederman (Eds.), *Scientific inquiry and nature of science*. Science & Technology Education Library, vol 25. Dordrecht: Springer.

van Uum, M. S., Verhoeff, R. P., & Peeters, M. (2016). Inquiry-based science education: Towards a pedagogical framework for primary school teachers. *International Journal of Science Education*, 38(3), 450–469.

Velayutham, S., Aldridge, J., & Fraser, B. (2011). Development and validation of an instrument to measure students' motivation and self-regulation in science learning. *International Journal of Science Education*, 33(15), 2159–2179.

White, B. Y., & Frederiksen, J. R. (1998). Inquiry, modeling, and metacognition: Making science accessible to all students. *Cognition and Instruction*, 16, 3–118.

Chapter 2

The use of easily accessible resources during hands-on practical activities in rural under-resourced Namibian schools

Eva Asheela, Kenneth Mlungisi Ngcoza and Joyce Sewry

Introduction

The advent of the new curriculum in Namibia in 1990 brought hope to many science teachers. For instance, central to the curriculum is learner-centredness whereby teachers are regarded as facilitators as opposed to being transmitters of knowledge (Namibian National Curriculum of Basic Education (NCBE), 2010; Nyambe, 2008). That is, the curriculum emphasises that learners should be central in the learning process. Furthermore, the NCBE stipulates that some effort should be made to contextualise and make science relevant to learners' everyday lives. It is also in favour of the use of easily accessible resources from the local environment (Asheela, 2017; Asheela, Ngcoza & Enghono, 2015; Ndevahoma, 2019; Shinana, 2019).

Despite the NCBE's ideals, however, it was found that there seemed to be tensions between curriculum formulation and implementation (Nyambe, 2008; Nyambe & Wilmot, 2012). For instance, there were a number of challenges which hindered the effective enactment of Learner-Centred Education (LCE), namely, teachers' self-doubt and lack of professional competence, inadequate or lack of professional support and inadequate academic background. Additionally, the reason most cited amongst many teachers is that there is a lack of resources to do hands-on and minds-on practical activities, especially in under-resourced rural schools (Heeralal, 2014), resulting in reluctance or a decrease in self-efficacy and motivation to include these in science lessons. For Bandura (1994, p.1), "perceived self-efficacy is defined as people's beliefs about their capabilities to produce designated levels of performance that exercise influence over events that affect their lives". Concurring, Yükseland Alci (2012) posit that teachers' actions and behaviours are related to their levels of self-efficacy. Embedded in self-efficacy is also motivation which, according to Cetin-Dindar andGeban (2017), is essential for active participation in the learning process.

According to Nyambe and Wilmot (2012), the above-mentioned constraining factors were exacerbated by the fact that teacher educators (lecturers) themselves were grappling with how to implement LCE. It is against this background, therefore, that this study sought to address three obstacles that seem to hinder practice: a) a perceived lack of resources; b) not knowing how quality hands-on and minds-on practical activities might be designed and delivered; and c) motivation and the self-efficacy that science teachers can in fact implement hands-on and minds-on practical activities after having seen and being part of such model lessons in a collaborative learning setting. The study thus sought to address the following research questions:

How does an intervention modelling science lessons using easily accessible resources in carrying out hands-on and minds-on practical activities:

1   Reduce in-service science teachers' reluctance to do hands-on and minds-on practical activities in their science classrooms; and
2   Increase their self-efficacy and motivation to include hands-on and minds-on practical activities in their science lessons?

Literature review

Hands-on practical activities have an important role to play in the science classroom (Abrahams & Millar, 2008; Jokiranta, 2014; Maselwa & Ngcoza, 2003; Millar, 2010; Ndevahoma, 2019). However, Maselwa and Ngcoza (2003), taking heed of Hodson's (1990) caution regarding practical activities – that they are usually of a cookbook or recipe approach – advise that the focus should be on sense-making and hence conceptual learning during hands-on practical activities. For Weick and Sutcliffe (2005), sense-making involves efforts to make science concepts comprehensible and understandable. Concurring, Cetin-Dindar and Geban (2017) emphasise that conceptual learning entails making meaningful connections of science concepts as opposed to rote learning.

To achieve this, Maselwa and Ngcoza (2003) recommend the 'predict-explain-explore-observe-explain' (PEEOE) approach to be employed during hands-on and minds-on practical activities. In this approach, they believe that learners or teachers are afforded an opportunity to make predictions before doing the practical activities and provide explanations thereof. The PEEOE approach is based on the constructivist perspective, central to which is learning in a social context (Vygotsky, 1978). Furthermore, these scholars encourage the use of easily accessible resources to bridge the gap between home or community and school science (Shinana, 2019). Oloruntegbe and Ikpe (2011) posit that teachers should consciously help learners during their science lessons to bridge the divide between *school science* and *household activities*. Similarly to Maselwa and Ngcoza (2003), these scholars hold that if learners are not exposed to purposeful hands-on and minds-on practical activities, they may not be able to relate science to the life-world at home. As far as they are concerned, a broad array of household activities are a good source of meaningful, experiential-based teaching and learning, affording learners to think both retroductively and retrodictively (Chikamori, Tanimura & Ueno, 2019). That is, learners are enabled to consider what they have learnt in the past to help them understand the present and the future.

In light of this, we argue that easily accessible resources could be in the form of the indigenous practice (Ogunniyi & Hewson, 2008; Ogunniyi & Ogawa, 2008), for example, making *oshikundu* [1] (Nikodemus, 2017; Shinana, 2019). In his study, for instance, Nikodemus used *oshikundu* to mediate learning of rates of reactions and associated concepts such as fermentation. On the other hand, Shinana worked *with* science teachers (Ngcoza & Southwood, 2015) to mobilise *oshikundu* to mediate learning of enzymes as well as promotion of inquiry-based approaches. In both these studies, the findings revealed that science was contextualised, resulting in greater conceptual learning and hence sense-making of the key scientific concepts developed.

A number of studies on the use of easily accessible resources to do hands-on and minds-on practical activities have also been conducted in South Africa (e.g. Kuhlane, 2011; Maselwa & Ngcoza, 2003). For example, Maselwa and Ngcoza used easily accessible resources such as old transparencies, scissors, rulers and plastic gloves to mediate learning of electrostatics. Conceptual learning was reinforced through developing mind maps and subsequently concept maps. In Erinosho's study conducted in Nigeria, learners and teachers were taken to local sites to observe activities related to what they learn at school, namely, metal production, food and textiles (Erinosho, 2013). The study revealed that learners were excited to be at the sites and were able to find relevance between what they studied in their classrooms and their local contexts.

From the aforementioned studies, it can be deduced that using easily accessible resources would be appropriate to create culturally sensitive or relevant pedagogies (Mhakure & Otulaja, 2017). As espoused by Vygotsky (1978), when easily accessible resources are used to mediate learning, this deepens notions of inclusivity, resulting in active participation by learners (Sedlacek & Sedova, 2017). That is, it has the potential to reinforce social interactions, which are essential to enhancing meaningful learning. It is precisely for these reasons that Mavuru and Ramnarain (2017) advocate taking into consideration learners' diverse socio-cultural backgrounds.

Theoretical frameworks: Socio-cultural theory

Central to Vygotsky's (1978) socio-cultural theory is the understanding that social interactions stimulate learning and hence are fundamental in the development of cognition. That is, learning takes place best in a social context. Building on the seminal work of Vygotsky, McRobbie and Tobin (1997) accentuate that social and personal aspects are indeed essential for learning to occur. Arguably, affordances for learning can be enhanced through hands-on and minds-on practical activities. Within the socio-cultural perspective, the concepts of *mediation, social interactions* and the *zone of proximal development (ZPD)* were used in this study. Vygotsky posits that mediation is essential in studying social processes since learning originates in social mediation. Furthermore, learning is realised through collaborative social interactions (Gibbons, 2003). In the context of this study, social interactions were stimulated using the hands-on and minds-on practical activities. Moreover, the concept of ZPD was appropriate as the science teachers involved were understandable at different zones of understanding. Stott (2016), however, cautions that there is no ZPD before learning activity takes place. Hence, she refers to the ZPD as the zone of proximal learning (ZPL) with an emphasis on *learning* rather than *development.* Indeed, Vygotsky posits that learning precedes development, which encompasses higher mental functions. Vygotsky further contends that the ZPD is created through and during social interactions. In the context of this study, social interactions and their enhanced learning features were included in the design of the intervention involving a group of teachers.

Research design and methodology

The design of the intervention sought to model science lessons using easily accessible resources in carrying out hands-on and minds-on practical activities so as to reduce in-service science teachers' reluctance and/or increase their motivation and self-efficacy to include practical activities in their science lessons. The focus was to address three obstacles that seem to be relevant to their motivation: a) a perceived lack of resources; b) not knowing how quality practical science lessons might be designed and delivered; and c) self-efficacy that they can in fact implement hands-on and minds-on practical activities after having seen and been part of such model lessons in a collaborative learning setting. The quality of the science lesson was of particular importance in the design with the PEEOE model chosen as a central feature of the design approach[2] (Maselwa & Ngcoza, 2003). Essentially, the first author co-designed and adapted activities with her co-researcher[3] on practical activities using easily accessible resources to render the interpretive paradigm relevant in this study.

Central to the interpretive paradigm is understanding human experiences from within the individual's intentional behaviours and actions (Bertram & Christiansen, 2015; Cohen, Manion & Morrison, 2018). Within the interpretive paradigm, a qualitative case study approach was employed (Maree, 2011; Merriam, 2009), the case being a week-long workshop of hands-on and minds-on practical science activities being attended by the participating in-service teachers. Data gathering techniques such as workshop discussions, observations and interviews are typical techniques associated with the interpretive paradigm and thus were suitable methods to collect data in order to answer the research question.

Research site, participants and sampling

In the context of this study, a combination of purposive and convenience sampling strategies were used (Cohen et al., 2018; Neuman, 2011). We used purposive sampling in the sense that all the in-service teachers in the study were at the time of this study doing the BEd (Honours) Science elective as part of their course in Namibia. We were looking at doing research *with* them rather *on* them, as reiterated by Ngcoza and Southwood (2015). Sampling was also convenient because both the in-service teachers in the sample and the first and second authors were attending the Honours and Masters contact sessions at the National Institute for Educational Development (NIED)[4] in Okahandja, Namibia. The first author was a Masters student and the second and third authors were co-supervisors. As a result, it was easy for the researchers to have access to the participants in the study (the in-service teachers) in order to conduct intervention workshops at one site.

The research participants were 21 in-service teachers who were students in a BEd (Honours) programme at a university in Eastern Cape, South Africa. The teachers were teaching at different rural schools in Namibia. Of the 21 participants, 10 were teaching Physical Science (also Mathematics in some cases) with experience ranging from 1 to 10 years, with an average of 5.5 years' teaching experience in Physical Science. Nine of the participants taught one of the Life Science subjects such as Biology, Natural Science and Health Education and Agriculture. Two of the participants taught only Mathematics although they had also specialised in Science at tertiary level.

Data gathering techniques

To gather data in this study, the following data gathering techniques were used: observations and workshop discussions, worksheets and mind maps, semi-structured interviews and reflections. We now discuss each of these below.

Observations and workshop discussions

The first author presented the workshops while the second author observed the contributions and social interactions during the intervention lessons. The workshop lessons were conducted over four consecutive days. The sessions ranged from 1 hour 30 minutes to 2 hours per session per day. During the workshops, the participants worked in groups of five or six and carried out hands-on practical activities using easily accessible resources, following the modelling and instructions of the workshop facilitator. Additionally, participants had to complete worksheets individually, and then as a group make predictions of the outcomes of the activities as well as provide explanations for their predictions. We observed that during the hands-on practical activities the science teachers discussed the activities in their local language Oshiwambo, which enabled meaningful social interactions. That is, the local language was used a resource rather than a barrier (Gibbons, 2003; Mavuru & Ramnarain, 2019; Msimanga & Lelliot, 2014).

The participants were also required to make observations and give explanations for their observations of the practical activities and thereafter produce mind maps and concept maps of the key science concepts which emerged from the activities. The intention was to surface and reinforce the relevant science concepts of the particular practical activity. Each group then presented and explained their concept mind maps of the activities. The participating teachers showed a great level of enthusiastic commitment – they actively helped to set up the apparatus and performed the practical activities (Sedlacek & Sedova, 2017). Table 2.1 summarises the activities conducted during the workshops.

Nine different hands-on practical activities (see Table 2.1) using easily accessible resources were carried out during the workshops. During the hands-on practical activities, an MEd student videotaped all the lessons. Also, a critical friend, who was a co-researcher, took some field notes while the first author facilitated during the workshops.

Semi-structured interviews and reflections

A semi-structured interview usually requires a participant to answer a set of pre-determined questions, and it also allows for probing and clarification of answers from the interviewee (Cohen et al., 2018; Maree, 2011). At the end of the workshops, the first author conducted semi-structured interviews with three participants who volunteered to be interviewed. The main aim of the interviews was to obtain in-depth information on the participants' experiences of the hands-on practical activities, what they enjoyed most and what they found to be challenging during the intervention. Furthermore, she also wanted to know if they felt that these activities were worth implementing in their own classroom situations or not. Additionally, all the teachers were requested to reflect on their experiences of the intervention.

Data analysis

The workshop videos were watched three times and narrative stories from the videos were written (Nhase, 2019; Sedlacek & Sedova, 2017). Thereafter, from the narrative stories some themes were identified. Similarly, the qualitative data from the semi-structured interviews and teachers' reflections were colour coded and categorised into sub-themes, and similar sub-themes were combined into themes (Creswell, 2012, 2014). Concepts from Vygotsky's (1978) socio-cultural theory such as *mediation, social interactions* and *zone of proximal development (ZPD)* were used as lenses to identify the themes. That is, a thematic and inductive approach to data analysis was employed (Cohen et al., 2018). From the analysed data, three themes emerged, namely:

Table 2.1 Summary of practical activities conducted during the workshops

Activities conducted	Easily accessible resources used
Day 1 • Preparation of the traditionally brewed beverage called *Ontaku/Oshikundu* • Preparation of yeast and sugar solution • Eggs in different liquids	• *Omahangu* flour • Flour from *Omahangu*/Sorghum germinated seeds • Residue from already fermented *Ontaku/Oshikundu* called *Oshipithitho* • Hot water (just below boiling point) • Cold water (at room temperature) • Bucket • Plastic bottles × 4 • Balloons × 4 • A plastic bottle (preferably 2 litre) • Lukewarm water • Yeast sachet • White sugar • Brown sugar • About 6 eggs per group • Vinegar • Lemon juice • Coca cola • 6 × beakers • Distilled water
Day 2 • Testing for carbon dioxide gas prepared in five different ways using lime water	• From the *Ontaku/Oshikundu* practical activity (from Day 1) • From yeast and sugar solution (from Day 1) • From exhaled air • From reaction of vinegar and bicarbonate of soda ($NaHCO_3$) • From reaction of egg shells and hydrochloric acid
Day 3 • Observations of eggs in different liquids (from Day 1) • Preparation of eggs (from Day 1) and potatoes for osmosis	(Finishing off tasks from day 1) • About 6 eggs per group • Vinegar • Lemon juice • Coca cola • 6 × beakers • Distilled water • Concentrated salt solution (salt sol.) • Concentrated sugar solution (sugar sol.) • Distilled water (water)
Day 4 • Preparation of hydrogen gas • Tea bags activity • Squashing can activity	• Caustic soda (sodium hydroxide) • Aluminium foil • Balloons • Plastic bottles • Tea bags • Hot water (boiling point) • Cold water (preferably from the refrigerator) • Cool drink cans (preferably 340 ml or bigger) • A source of heat (methylated spirits) • A cold water bath

- Overcoming constraints – Potential of easily accessible resources;
- Overcoming constraints – Motivation to teach hands-on practical; and
- Learning potential – Self-efficacy and hence seeing the power of practical activities as a learning affordance.

Validity, trustworthiness and reliability

Validity, trustworthiness and reliability of this study were strengthened by the use of multiple data generation techniques (Maree, 2011), namely: worksheets, mind maps, workshop behaviours and discussions, observations, workshop videos and transcripts, interviews and written reflections of participants. Furthermore, the transcripts from the semi-structured interviews were given back to the interviewees for member checking (Cohen et al., 2018). In addition the input and comments of a variety of players examining and commenting on the data were available, e.g. the co-researcher, the two supervisors, another Masters student, and discussions with other academics.

Ethical considerations

All the science teachers involved in this study gave informed consent to participate. However, one of them reflected that:

> First and for most [sic], I would like to state that when I heard that we were requested to be part of our colleagues' research projects, my concern was that we were forced to help these students to complete their projects in one way or another because of their seniority. Secondly, I felt that it was time consuming as our programme was even extended to the evening sessions of which it was not the case in the first year. I also felt that this was a double task for me and it would not allow me to complete my own tasks on time. However, after getting involved, I learned that most of my problems that I had and doubts on some parts of my research presentation were automatically answered and misconceptions were cleared. I also realised that getting myself involved gave me more insights and got to acknowledge that the colleagues were actually not researching on me/us but rather they were researching with me/us.

From this excerpt, it is evident that positionality or power gradients need to be addressed appropriately in research. Ngcoza and Southwood (2015) emphasise the importance of doing research *with* participants rather than *on* them; participants should be treated with utmost respect throughout the research process. Notwithstanding the positionality of the researcher, it seems a good rapport was established and there was mutual benefit for both the participants and the researcher.

Results and discussion

The data generated during the intervention through workshop discussions, observations, semi-structured interviews and reflections aimed at addressing the research question:

> How does an intervention modelling science lessons using easily accessible resources in carrying out hands-on practical activities:
>
> 1   Reduce in-service science teachers' reluctance to do hands-on practical activities in their science classrooms; and
> 2   Increase their self-efficacy and motivation to include hands-on practical activities in their science lessons?

Overcoming constraints: Potential of easily accessible resources

It was confirmed by this study that most of these science teachers do not conduct hands-on practical activities in their schools because of the lack of or unavailability of resources and laboratories in their schools. For example, three of these participants lamented that:

> "Many of us (teachers) fail to expose our learners to these activities claiming that there are no laboratories or chemicals at school." (RT5)
> "And moreover, we have less knowledge on how to access resources within our reach." (RT6)

In contrast, however, one participant commented that:

> "We have access to all the resources. It is only that we really do not know how to use them in terms of performing practical activities using such resources. Yes because teachers are unaware of those simple experiments we can carry out." (RT10)

This reflection sheds light on the fact that in some schools resources might be available but science teachers are unable to use them. Indeed, these reflections point to the need for continuing professional development or professional learning communities (Ngcoza, 2007; Ngcoza & Southwood, 2019) of science teachers, as reflected in the following comments:

> "...now I can be able to prepare practical activities even if the school does not have a science lab or the lab is not equipped." (RT4)
> "It encouraged me/gave me insight that although there is no lab at school, I can still use available/cheap materials at home/shops to do practical activities." (RT9)

Similarly, some participants reflected as follows in the semi-structured interviews.

> "Now I am remembering now the activity that we have done that one for caustic soda and the Magnesium, I mean the Aluminium foil... I didn't know that Aluminium can react like that." (IT1F)
>
> "This is wonderful for everyday material because at tertiary education, you are being taught to go and teach at different schools, and if you are only taught using regular materials, those materials you may not be having them at your school so using everyday material it helps you to prepare the gases like the one we were testing or to, it helps you to carry out different investigations regardless of the availability or unavailability." (IT2M)
>
> "Some of the things are the things that I even have at home, every day I see them some of them I deal with them, I eat them, eggs and Oshikundu things I drink them but then the way you put them up in the way of activities it was really something else for me... but now I never just thought of using a weak acid, or even a strong acid to look at now how the egg is being [sic] reacting with the acid... the foil, I used it, I threw it away, I know it is a metal, I talk about it that aluminium can be used for making food containers, foil is one of the examples, but then I never came to an example or idea of taking it as a metal, reacting it with something to produce that hydrogen..." (IT3F)

The participants acknowledged that most of the materials used in the workshops were readily available, within their reach and affordable. This is demonstrated by the following quotes:

> "Also, it is cheaper as the resources/materials to be used in the practical activities can be generated from the environment." (RT8)
>
> "Yes because we used to give excuses that there are [sic] no equipment at school but now I have an idea that you can get them in shops at a cheaper price or use local materials." (RT13)
>
> "...materials or resources does [sic] not cost too much to buy them." (RT17)

These findings support the literature that suggests that in order for teachers to use resources, they must have knowledge of the resources. For instance, Shulman (1986, 1987) states that in order for teachers to be able to effectively mediate learning, they should know the best strategies to employ in order to make the content understandable to their learners. Scholars such as Aikenhead and Jegede (1999) and Mhakure and Otulaja (2017) reiterate that teachers need to develop culturally sensitive pedagogies in order to facilitate smooth border crossing between the learners' everyday knowledge and the science taught in schools. It is acknowledged, however, that in order for teachers to be able to use familiar teaching resources, they need to have the knowledge of those resources themselves.

Overcoming constraints: Motivation to now teach hands-on practical

The science teachers involved in this study stated that the demonstrated and modelled lessons and activities captured their interest and motivated them and gave them a belief that they can do this. The series of workshops over a week had stimulated questioning and thinking and enhanced active participation (Sedlacek & Sedova, 2017). These ideas are supported by the following quotes:

> "This was a great approach that I could not even use or think of during my teaching. It helped us engage with our thinking abilities." (RT2)
>
> "It was an excellent way of teaching, learners taught in this way would be given time to think about the aspect/concept before the experiment is done." (RT6)
>
> "It promotes learning and encourages active participation because learners are dealing with materials they are familiar with." (RT7)
>
> "...it is very enjoyable, the way I was myself in that workshop I could put myself as a learner... it was an exciting moment for me already... So already, even if I know it, it was a very exciting moment for me just to test that hydrogen coming from the foil, that I know about, that I see it coming from my kitchen." (IT3F)

In our view, these excerpts resonate with the views of scholars such as Hodson (1990), Maselwa and Ngcoza (2003) and Erinosho (2013) who state that hands-on practical activities are an enjoyable exercise that promote motivation. Similarly, in Ngcoza et al.'s (2016) South African case study, it emerged that learners who participated in science expos, which usually involve interesting hands-on practical experiments, become more motivated to study science at school. This is also in line with Vygotsky's (1978) socio-cultural theory, which emphasises the importance of context in teaching and learning. Atallah, Bryant & Dada (2010) indicated that real-life experiences promote interest and confidence, such as the participation in these hands-on practical activities using easily accessible resources in the context of this study.

Learning potential – Self-efficacy: Now seeing the power of practical activities as a learning affordance

The in-service teacher participants involved in this study felt that the hands-on practical activities were useful in enhancing their understanding of their learners, having experienced first the modelled practical activity lessons. For example, three teachers reflected on the potential of practical activities in teaching their own learners:

"Doing practical activities helps learners to understand and internalise what they are being taught. It helps them to move away from memorising facts..." (RT2)

"It is a powerful form of doing practical activities because it allows learners to get to understand that science is around them." (RT16)

"I feel practical activities should really be reinforced in classrooms. I feel that they make learning more meaningful and fun." (RT18)

The insights from the participants' reflections indicate that the hands-on practical activities conducted during the workshop had a positive influence on uncovering the science concepts in lessons for their learners. It could be hypothesised that this was in part the result of a number of elements of the lessons: the familiar materials used from the participants' everyday lives, the hands-on practical activities, and the PEEOE lesson design approach. Further elements were the group work and the extended nature of the experience over a week of lessons.

Roschelle (1995) and Stears, Malcolm and Kowlas (2003) emphasise that learners relate strongly to content that relates to their lives and such knowledge could be in the form of local or indigenous knowledge, as evidenced in this study. These ideas are supported by Mukwambo (2012) who reports that the integration of indigenous knowledge promotes teaching and learning scenarios in which both teachers and learners engage in knowledge construction.

Also, since the participants worked in groups during the practical activities, they were able to identify scientific concepts in relation to the practical activities they were engaged in. For instance, from the practical activity on the preparation of *Ontaku/Oshikundu*, the various groups were able to come up with the different concepts.

Table 2.2 shows collated scientific concepts which emerged from the mind maps of the five groups.

Essentially, the hands-on practical activities enable learners to make links between concepts, and also provide an opportunity to apply their scientific understanding to the real-world application of science (Millar, 2010; Woodley, 2009). These findings are also in line with the findings of a number of South African researchers who also conducted case studies using easily accessible resources to conduct practical activities in the science classrooms (e.g. Kuhlane, 2011; Maselwa & Ngcoza, 2003; O'Donoghue et al., 2007). It emerged in these studies that knowledge of resources and how to use them was pivotal.

Conclusion and implications

In initially conceiving and planning this research, we were aware of the resistance of many science teachers in Namibia to the inclusion of many practical activities in their science lessons. We sought to explore further the nature of this resistance and especially ways to possibly overcome the real or perceived

Table 2.2 Concepts that emerged from the Ontaku/Oshikundu practical activity

Group 1	Group 2	Group 3	Group 4	Group 5
Alcohol	Anaerobic	Anaerobic	Alcohol	Bubbles
Carbon dioxide (CO_2)	Carbon dioxide (CO_2)	Alcohol	Anaerobic respiration	Carbon dioxide (CO_2)
Catalyst	Catalyst	Catalyst	Biological catalyst	Catalyst
Energy	Expansion	Chemical	Carbon dioxide (CO_2)	Decanting
Fermentation	Fermentation	Carbon dioxide (CO_2)	Enzymes	Diffusion
Floating	Inflated	Density	Expansion	Expansion
H_2O	Molecules	Enzyme	Fermentation	Fermentation
Glucose	Observation	Expand	Inflation	Floating
Synthesis	Particles	Fermentation	Product	Gas pressure
	Products	Filtration	Reactants	Sinking
	Reactants	Gas		Stretching
	Reaction	Inflate		Reaction
	Temperature	Mixture		Residue
		Reactants		Temperature
		Reaction		
		Residue		
		Respiration		
		Temperature		

obstacles. Asheela (2017) had just established that Namibia teachers were grappling with hands-on practical activities in their science classes and uncovered that the lack of resources was one of the perceived main obstacles stated by teachers, but she also uncovered that many teachers felt that they just did not know how to design and run hands-on practical activity science lessons.

The design of the intervention sought to address these two primary obstacles directly: resources and what quality practical activity science lessons might look like. But the aim was also to give the in-service teacher participants enough time to experience, internalise and reflect on the suggested lesson approach and build the self-belief that they too can and should follow it themselves.

The findings from the sample of in-service teachers attending a part-time BEd Honours course in Namibia who participated in a week-long series of workshops modelling lessons were:

1. Overcoming constraints: Potential of easily accessible resources

After being exposed to examples and the modelled lessons using easily accessible resources, the in-service teachers saw the potential of their use not only as an alternative if they did not have the normal science materials, but also as being of value and interest for their own science lessons.

2. Overcoming constraints: Motivation to teach hands-on practical science

The modelling of practical activities science lessons with the active participation of the in-service science teachers as 'the learners' resulted in bringing about a change in their motivation in wanting to try these lessons in their own classrooms.

The combined effect of the main design components of the intervention is probably responsible for the outcome: the modelling of quality lessons, the use of easily accessible resources and rich interactions, plus the intellectual interest and challenge of the PEEOE model.

3. Learning potential: Self-efficacy – Now seeing the power of practical activities as a learning affordance

Their own experience of the fun and enthusiasm experienced during these lessons, plus the other elements of the lessons, such as the PEEOE approach and identifying the relevant science concepts of an activity, had the effect of persuading the in-service teachers of the potential of practical activity lessons, enhancing the learning and understanding of science of their own learners.

In summary, the very positive and encouraging results, of this relatively simple design has implications for the design and delivery of continuing professional development for science teachers.

Notes

1 *Oshikundu* or *Ontaku* is a non-alcoholic nutritious beverage made by many families in Namibia.
2 The design of the practical activities science workshop using easily accessible resources was modelled on a workshop run by Mr Richard Grant, retired lecturer of Physics at Rhodes University, attended a year earlier by the first and second authors and modified with the inclusion of different science experiments and the prediction/ explain model and accompanying activities.
3 The co-researcher in this study focussed on assessment for learning during hands-on practical activities.
4 It is at this site where the contact sessions for the Namibian BEd and MEd programs are conducted.

References

Abrahams, I., & Millar, R. (2008). Does practical work really work? A study of the effectiveness of practical work as a teaching and learning method in school science. *International Journal of Science Education*, 30(14), 1945–1969.

Aikenhead, G. S., & Jegede, O. J. (1999). Cross-cultural science education: A cognitive explanation for a cultural phenomenon. *Journal of Research in Science Teaching*, 36(3), 269–287.

Asheela, E. N. (2017). *An intervention on how easily accessible resources to carry out hands-on practical activities in science influences science teachers' conceptual development and dispositions.* Unpublished master's thesis, Education Department, Rhodes University, Grahamstown.

Asheela, E. N., Ngcoza, M. N., & Enghono, A. (2015). An indigenous practice of making a traditional beverage called *Oshikundu* as a strategy to enhance conceptual understanding. In M. B. Ogunniyi & K. R. Langenhoven (Eds.), *The relevance of indigenous knowledge to African socioeconomic development in the 21st century* (pp. 282–289). African Association for the Study of Indigenous Knowledge Systems. Windhoek: UNAM.

Atallah, F., Bryant, S. L., & Dada, R. (2010). Learners' and teachers' conceptions and dispositions of mathematics from a Middle Eastern perspective. *US-China Education Review*, 7(8), 43–49.

Bandura, A. (1994). Self-efficacy. In V. S. Ramachaudran (Ed.), *Encyclopedia of human behaviour* (Vol. 4, pp. 71–81). New York: Academic Press.

Bertram, C., & Christiansen, I. (2015). *Understanding research: An introduction to reading research.* Pretoria: Van Schaik Publishers.

Cetin-Dindar, A., & Geban, O. (2017). Conceptual understanding of acids and bases concepts and motivation to learn chemistry. *The Journal of Educational Research*, 110(1), 85–97. doi:10.1080/00220671.2015.1039422

Chikamori, K., Tanimura, C., & Ueno, M. (2019). Transformational model of education for sustainable development (TMESD) as a learning process of socialization. *Journal of Critical Realism.* doi:10.1080/14767430.2019.1667090

Cohen, L., Manion, L., & Morrison, K. (2018). *Research methods in education* (8th ed.). London: Routledge.

Creswell, J. W. (2012). *Educational research: Planning, conducting, and evaluating quantitative and qualitative research* (4th ed.). Boston, MA: Pearson Education.

Creswell, J. W. (2014). *Research design: Qualitative, quantitative, and mixed methods* (4th ed.). Los Angeles: Sage Publications.

Erinosho, S. Y. (2013). Integrating indigenous science with school science for enhanced learning: A Nigerian example. *International Journal for Cross-Disciplinary Subjects in Education (IJCDSE)*, 4(2), 1137–1143.

Gibbons, P. (2003). Mediating language learning: Teacher interactions with ESL students in a content-based classroom. *TESOL Quarterly*, 37(2), 247–273.

Heeralal, P. J. H. (2014). Barriers experienced by Natural Sciences teachers in doing practical work in primary schools in Gauteng. *International Journal of Science Education*, 7(3), 795–800.

Hodson, D. (1990). A critical look at practical work in school science. *School Science Review*, 70(256), 33–40.

Jokiranta, K. (2014). *The effectiveness of practical work in science education*. Unpublished bachelor's thesis, University of Jyväskylä, Jyväskylä.

Kuhlane, Z. (2011). *An investigation into the benefits of integrating learners' prior everyday knowledge and experiences during teaching and learning of acids and bases in grade 7: A case study*. Unpublished master's thesis, Education Department, Rhodes University, Grahamstown.

Maree, K. (Ed.). (2011). *First steps in research*. Pretoria: Van Schaik Publishers.

Maselwa, M. R., & Ngcoza, K. M. (2003). 'Hands-on', 'minds-on' and 'words-on' practical activities in electrostatics: Towards conceptual understanding. In D. Fisher & T. Marsh (Eds), *Making science, mathematics and technology education accessible to all: Proceedings of the Third International Conference on Science, Mathematics and Technology Education* (pp. 649–659). Perth: Key Centre for School Science and Mathematics.

Mavuru, L., & Ramnarain, U. (2017). Teachers' knowledge and views on the use of learners' socio-cultural background in teaching Natural Sciences in Grade 9 township classes. *African Journal of Research in Mathematics, Science and Technology Education*, 21(2), 176–186.

Mavuru, L., & Ramnarain, U. (2019). Language affordances and pedagogical challenges in multilingual grade 9 Natural Sciences classrooms. *African Journal of Research in Mathematics, Science and Technology Education*, 21(2), 176–186.

McRobbie, C., & Tobin, K. (1997). A social constructivist perspectives on learning environments. *International Journal of Science Education*, 19(2), 193–208.

Merriam, S. B. (2009). *Qualitative research: A guide to design and implementation* (2nd ed.). San Francisco: Jossey-Bass.

Mhakure, D., & Otulaja, F. S. (2017). Culturally-responsive pedagogy in science education: Narrative of the divide between indigenous and scientific knowledge. In F. S. Otulaja & M. B. Ogunniyi (Eds.), *The world of science education: Handbook of research in science education in sub-Saharan Africa* (pp. 81–100). Rotterdam: Sense Publishers.

Millar, R. (2010). *Developing students' understanding of science: The role of practical work*. Metodelab conference, lecture notes, University of York, Department of Education; Copenhagen.

Msimanga, A., & Lelliot, A. (2014). Talking science in multicultural contexts in South Africa: Possibilities and challenges for engagement in learners' home language in high school classrooms. *International Journal of Science Education*, 36(7), 1159–1183.

Mukwambo, M. (2012). *Understanding trainee teachers' engagement with prior everyday knowledge and experiences in teaching Physical Science concepts: A case study*. Unpublished master's thesis, Education Department, Rhodes University, Grahamstown.

Namibia Ministry of Education. (2010). *The National Curriculum for Basic Education (NCBE)*. Okahandja: NIED.

Ndevahoma, M. K. (2019). *Enactment of hands-on practical activities through using easily accessible resources in a Grade 10 Physical Science classroom.* Unpublished master's thesis, Education Department, Rhodes University, Grahamstown.

Neuman, L. W. (2011). *Social research methods: Qualitative and quantitative approaches* (7th ed.) Boston, MA: Pearson.

Ngcoza, K. M. (2007). *Science teachers' transformative and continuous professional development: A journey towards capacity-building and reflexive practice.* Unpublished doctoral thesis, Education Department, Rhodes University, Grahamstown.

Ngcoza, K. M., & Southwood, S. (2015). Professional development networks: From transmission to co-construction. *Perspectives in Education*, 33(1), 4–14.

Ngcoza, K., & Southwood, S. (2019). Webs of development: Professional networks as spaces for learning. *Pythagoras, 40*(1), a409. http://doi.org/10.4102/pythagoras.v40i1.409

Ngcoza, K. M., Sewry, J., Chikunda, C., & Kahenge, W. (2016). Stakeholders' perceptions of participation in science expos: A South African case study. *African Journal of Research in Mathematics, Science and Technology Education*, 20(2), 189–199.

Nhase, Z. (2019). *An exploration of how Grade 3 Foundation Phase teachers develop basic scientific process skills using an inquiry-based approach in their classrooms* Unpublished doctoral study, Education Department, Rhodes University, Grahamstown.

Nikodemus, K. S. (2017). Exploring how Grade 11 Physical Science learners make sense of concepts on rates of reactions through the inclusion of indigenous practice of making Oshikundu: A Namibian case study. In M. B. Ogunniyi & K. R. Langenhoven (Eds.), *The relevance of indigenous knowledge to African socioeconomic development in the 21st century.*(pp. 7–16). African Association for the Study of Indigenous Knowledge Systems. Windhoek: UNAM.

Nyambe, J. K. (2008). *Teachers' interpretation of learner-centred education pedagogy: A case study.* Unpublished doctoral thesis, Education Department, Rhodes University, Grahamstown.

Nyambe, J. K., & Wilmot, D. (2012). New pedagogy, old pedagogic structures: A fork-tongued discourse in Namibian teacher education reform. *Journal of Education, 55*, 55–82.

O'Donoghue, R., Lotz-Sisitka, H., Asafo-Adjei, R., Kota, L., & Hanisi, N. (2007). Exploring learning interactions arising in school-in-community context of socio-ecological risk. In A. E. J. Walls (Ed.), *Social learning towards a sustainable world* (pp. 435–447). Wageningen, The Netherlands: Wageningen Academic Publishers.

Ogunniyi, M. B., & Hewson, M. G. (2008). Effect of an argumentation-based course on teachers' disposition towards a science-indigenous knowledge curriculum. *International Journal of Environmental & Science Education*, 3(4), 159–177.

Ogunniyi, M. B., & Ogawa, M. (2008).The prospects and challenges of training South African and Japanese educators to enact an indigenised science curriculum. *South Africa Journal of Higher Education*, 22(1), 175–190.

Oloruntegbe, K. O., & Ikpe, A. (2011). Ecocultural factors in students' ability to relate science concepts learned at school and experienced at home: Implications for chemistry education. *Journal of Chemical Education, 88*(3), 266–287.

Roschelle, J. (1995). *Learning in interactive environments: Prior knowledge and new experience.* In J. H. Falk & C. D. Dierking (Eds.), *Public institutions for personal learning: Establishing a research agenda* (pp. 37–51). Washington, DC: American Association of Museums.

Sedlacek, M., & Sedova, K. (2017). How many are talking? The role of collectivity in dialogic teaching. *International Journal of Educational Research*, 85, 99–108.

Shinana, E. N. L. (2019). *Mobilising the indigenous practice of making Oshikundu using inquiry-based approach to support Grade 8 Life Science teachers in mediating learning of enzymes*. Unpublished master's thesis, Education Department, Rhodes University, Grahamstown.

Shulman, L. S. (1986). Those who understand: Knowledge growth in teaching. *Educational researcher*, 15(2), 4–14.

Shulman, L. S. (1987). Knowledge and teaching: Foundations of new reform. *Harvard Educational Review*, 57(1), 1–21.

Stears, M., Malcolm, C., & Kowlas, L. (2003). Making use of everyday knowledge in the science classroom. *African Journal of Research in Mathematics, Science and Technology Education*, 7(1), 109–118.

Stott, D. (2016). Making sense of the ZPD: An organising framework for mathematics education research. *African Journal of Research in Mathematics, Science and Technology Education*, 20(1), 25–34.

Vygotsky, L. S. (1978). *Mind in society: Interaction between learning and development*. Cambridge, MA: Harvard University Press.

Weick, K. E., & Sutcliffe, K. M. (2005). Organising and the process of sense-making. *Organization Science*, 16(4), 409–421.

Woodley, E. (2009). Practical work in school science: Why is it important? *School Science Review*, 91(335), 49–51.

Yüksel, G., & Alci, B. (2012). Self-efficacy and critical thinking dispositions as predictors of success in school practicum. *International Online Journal of Educational Sciences*, 4(1), 81–90.

Chapter 3

Advances in inquiry-based science education in Zimbabwean schools

Lydia Mavuru and Washington T. Dudu

Introduction

Millar (2004) defines 'practical work' as any teaching and learning activity which involves the learners, at some point, observing or manipulating real objects and materials. Learners can observe or manipulate such objects anywhere, be it the school laboratory or in any out-of-school setting (Millar, 2004). In order for any practical work to be meaningful to learners, it should be designed in such a way that it stimulates learners' thinking (Millar, 2004). Inquiry-based learning is an approach that aims at nurturing the thinking, reflection and problem-solving capabilities among science learners. During inquiry-based science engagements, learners identify problems, generate research questions, design and conduct investigations, and formulate, communicate and defend hypotheses, models and explanations (Abd-El-Khalick et al., 2004). Accordingly, as early as 2000 the National Research Council (NRC) identified features of inquiry-based teaching. These include involving learners in: asking scientifically oriented questions; focusing on providing or finding evidence; evaluating the explanations they give in relation to alternative views; and communicating and justifying the given explanations. Millar, Tiberghien and Le Maréchal (2002) developed a model (Figure 3.1) to demonstrate the role that practical work plays in linking what learners experience in real life to the ideas they develop during the science learning process, which forms the basis of acquiring scientific knowledge.

The essence of this model is that practical work facilitates learners' knowledge and understanding of science as they relate observation with learned information. It concretises abstract ideas that the learners hold about a

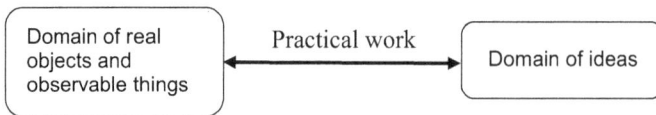

Figure 3.1 Linking of two domains of knowledge
(Adapted from Millar et al., 2002)

phenomenon. Previous studies have reported that inquiry-based learning experiences enhance learners' motivation to learn science (Crawford, 2012), improve understanding of concepts (Gott & Duggan, 2002), facilitate collaboration between learners (Hofstein & Lunetta, 2004), foster greater learner achievement (Edelson, 1998) and help develop processing skills (NRC, 1996).

In determining the inquiry-based practices encapsulated in the Zimbabwean science curriculum, Millar's (2009) Practical Activity Analysis Inventory (PAAI) is used as outlined in the section on inquiry-based practices encapsulated in the Zimbabwean science curriculum.

In this analysis we should be reminiscent of what Abrams, Southerland and Evans (2007) said about inquiry-based practices: that the degree of inquiry depends on who is responsible for each of the three key activities of asking questions, collecting data and interpreting data. This can be placed on a continuum from closed inquiry through guided inquiry to open-ended inquiry (Blanchard et al., 2010). Blanchard et al. (2010) categorised the first level (level zero) as closed inquiry, where learners confirm or verify known scientific principles by following a given procedure (normally referred to as cookbook laboratories). In "structured" inquiry (level 1), teachers provide learners with a question and a method and learners' only responsibility is to interpret results. In "guided" inquiry (level 2), learners determine the method of investigation and how to interpret the results. Blanchard et al. (2010) refer to open inquiry (level 3) as the highest level of inquiry, where teachers allow learners to develop their own questions and design their own investigations.

With this in mind, the following sections will analyse the Zimbabwean science curriculum and expectations for its implementation, review previous studies on how teachers implement inquiry-based practices, and analyse assessments and assessment reports to determine the extent to which inquiry-based practices are portrayed in the syllabi, classroom practical work and in the assessments, respectively. The analysis takes inquiry as both an instructional approach and a learning outcome. As such, the focus is both on how teachers are expected to engage learners in learning activities that are investigative in nature and how learning outcomes are achieved as evidenced in assessment and learner performance. Firstly, the context of the Zimbabwe science education system in high schools is presented in the next section.

Context of the Zimbabwe high school science education system

The current Zimbabwean high school science education curriculum is subdivided into four subjects: Combined Science, Biology, Chemistry and Physics. Combined Science is compulsory for all learners in the first two years of high school. Depending on the learners' ability and interest in continuing studying science, learners get to choose the other three specialisations. Alternatively, learners can continue to study Combined Science until fourth year.

Inquiry-based practices encapsulated in the Zimbabwean science curriculum

The investigative approach is the method used for the teaching and learning of high school science in Zimbabwe. The rationale is to develop independent learning, which is envisaged to continue long after high school years. Before speculating on what previous studies portray about the reality of inquiry-based practices implemented in Zimbabwean science classrooms, it is important to examine the extent to which the curriculum articulates these inquiry-based practices. In determining the inquiry-based practices encapsulated in the Zimbabwean science curriculum, a critical analysis will be provided of the aims, objectives and activities of the Ordinary Level (O-Level) and Advanced Level (A-Level) sciences.

It is believed that engaging learners in practical work enhances their understanding of scientific concepts (Millar, 2009). However, questions have been posed regarding whether learners acquire the intended knowledge and skills from the practical activities. Previous research has confirmed that the intentions of the practical activity are not always achieved, which then leaves scholars with critical questions on the contribution of practical work in science learning. A point to note is the manner in which the practical work is conceived and utilised in the science classroom, which Hodson (1996) labelled as ill-conceived and unproductively practised. Millar developed a tool to determine the effectiveness of practical work by systematically analysing and evaluating the objectives, design and presentation features. In order to determine the nature and level of the inquiry-based science education practised in Zimbabwean schools, this chapter considers the model of developing and evaluating the effectiveness of practical activities espoused by Millar, Tiberghien and Le Maréchal (2002), which is encapsulated in Millar's (2009) PAAI. The model highlights four stages: (1) Developer's intention for learning; (2) Activity specifications of what learners should do; (3) Classroom events indicating what learners actually did; and (4) Outcomes of what has been learned. Figures 3.2 and 3.3 provide the summary of the implementation of this model in analysing the Biology curriculum in terms of its scope, its intentions, the suggested activities and the stipulated classroom learning events.

Figure 3.2 shows the alignment of A-Level Biology objectives, activities and classroom learning events. It depicts the expected inquiry-based practices (curriculum objectives), possible activities that learners can engage with in order to ensure achievement (activity specifications), and how the teaching and learning process should take place in the science classroom (classroom teaching and learning events). The same analysis is made for O-Level Biology in Figure 3.3.

Tables 3.1 and 3.2 show the objectives for Chemistry and Physics practical work in A- and O-Levels, respectively. Similar objectives are set for Biology.

Curriculum stipulations: A-Level Biology

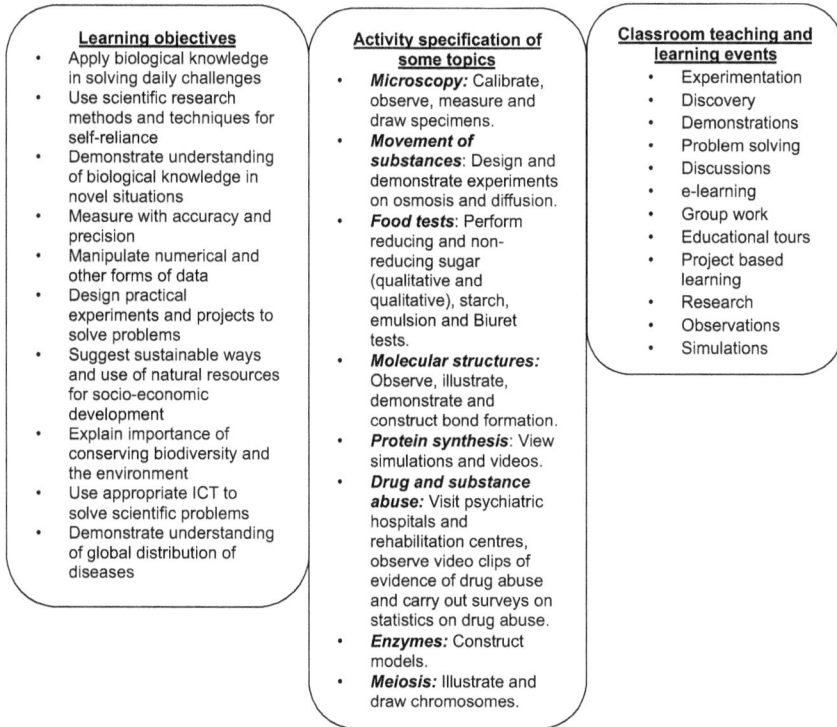

Learning objectives	Activity specification of some topics	Classroom teaching and learning events
• Apply biological knowledge in solving daily challenges • Use scientific research methods and techniques for self-reliance • Demonstrate understanding of biological knowledge in novel situations • Measure with accuracy and precision • Manipulate numerical and other forms of data • Design practical experiments and projects to solve problems • Suggest sustainable ways and use of natural resources for socio-economic development • Explain importance of conserving biodiversity and the environment • Use appropriate ICT to solve scientific problems • Demonstrate understanding of global distribution of diseases	• *Microscopy:* Calibrate, observe, measure and draw specimens. • *Movement of substances*: Design and demonstrate experiments on osmosis and diffusion. • *Food tests*: Perform reducing and non-reducing sugar (qualitative and qualitative), starch, emulsion and Biuret tests. • *Molecular structures:* Observe, illustrate, demonstrate and construct bond formation. • *Protein synthesis*: View simulations and videos. • *Drug and substance abuse:* Visit psychiatric hospitals and rehabilitation centres, observe video clips of evidence of drug abuse and carry out surveys on statistics on drug abuse. • *Enzymes:* Construct models. • *Meiosis:* Illustrate and draw chromosomes.	• Experimentation • Discovery • Demonstrations • Problem solving • Discussions • e-learning • Group work • Educational tours • Project based learning • Research • Observations • Simulations

Figure 3.2 Analysis of inquiry-based practices in A-Level Biology (Ministry of Primary and Secondary Education, 2015)

Both Tables 3.1 and 3.2 indicate the emphasis placed on planning, designing and making observations, as well as the interpretation of results and evaluation of the investigation process to enhance accuracy of results.

To ensure effective practical work in science classrooms, Millar (2009) designed the Practical Activity Analysis Inventory (PAAI) tool, which stipulates the importance of learning objectives for the practical activity, how the activity is designed and its presentation. PAAI prescribes four stages that teachers should use to develop and evaluate practical activities, which include learning intentions (objectives), practical activity intentions (activity specifications), learners' involvement (classroom events) and actual learning undertaken by learners (learning outcomes). Both Figures 3.2 and 3.3 reveal how the Biology syllabi have been organised according to the first three stages of Millar's PAAI. The PAAI framework also categorises learning outcomes into three groups. The first seeks to develop learners' knowledge and understanding of the natural world, which can be matched to the objectives: *Explain the importance of conserving*

Curriculum stipulations: O-Level Biology

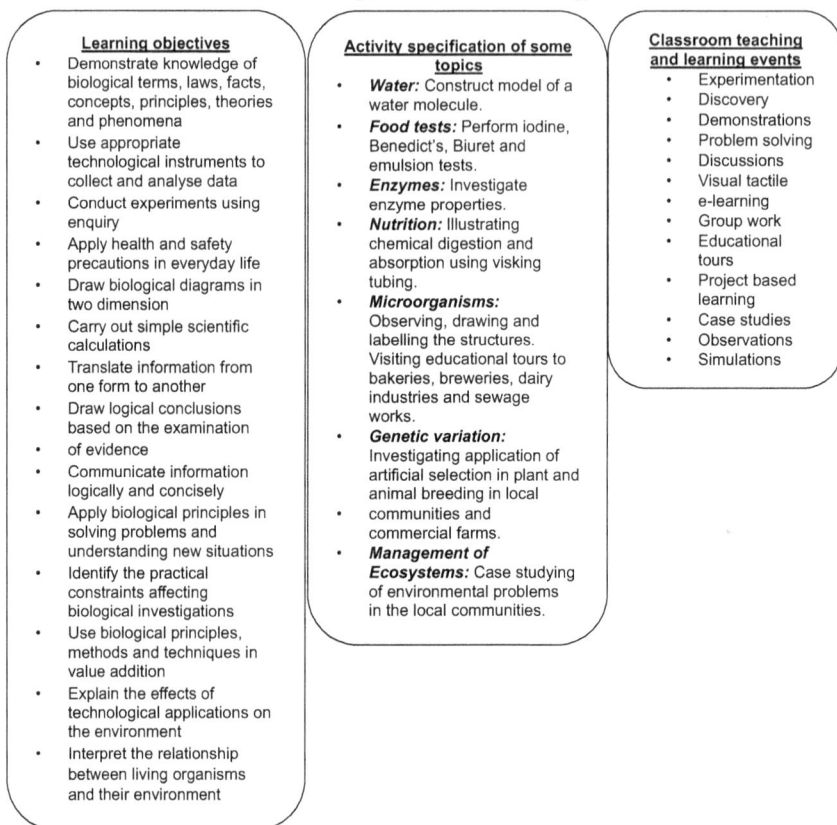

Learning objectives
- Demonstrate knowledge of biological terms, laws, facts, concepts, principles, theories and phenomena
- Use appropriate technological instruments to collect and analyse data
- Conduct experiments using enquiry
- Apply health and safety precautions in everyday life
- Draw biological diagrams in two dimension
- Carry out simple scientific calculations
- Translate information from one form to another
- Draw logical conclusions based on the examination
- of evidence
- Communicate information logically and concisely
- Apply biological principles in solving problems and understanding new situations
- Identify the practical constraints affecting biological investigations
- Use biological principles, methods and techniques in value addition
- Explain the effects of technological applications on the environment
- Interpret the relationship between living organisms and their environment

Activity specification of some topics
- *Water:* Construct model of a water molecule.
- *Food tests:* Perform iodine, Benedict's, Biuret and emulsion tests.
- *Enzymes:* Investigate enzyme properties.
- *Nutrition:* Illustrating chemical digestion and absorption using visking tubing.
- *Microorganisms:* Observing, drawing and labelling the structures. Visiting educational tours to bakeries, breweries, dairy industries and sewage works.
- *Genetic variation:* Investigating application of artificial selection in plant and animal breeding in local
- communities and commercial farms.
- *Management of Ecosystems:* Case studying of environmental problems in the local communities.

Classroom teaching and learning events
- Experimentation
- Discovery
- Demonstrations
- Problem solving
- Discussions
- Visual tactile
- e-learning
- Group work
- Educational tours
- Project based learning
- Case studies
- Observations
- Simulations

Figure 3.3 Analysis of inquiry-based practices in O-Level Biology (Ministry of Primary and Secondary Education, 2015)

Table 3.1 Analysis of inquiry-based practices in A-Level Chemistry and Physics

A-Level Chemistry	A-Level Physics
• Plan investigations; • Use techniques, apparatus and materials; • Make and record observations; • Interpret and evaluate observations and experimental results; • Select techniques, apparatus and materials; • Evaluate methods and suggest possible improvements; • Design a practical solution to a real-life problem using knowledge of Chemistry.	• Follow a detailed set or sequence of instructions and use techniques, apparatus and materials safely and effectively; • Make observations and measurements with due regard for precision and accuracy; • Interpret and evaluate observations and experimental data; • Identify a problem, design and plan investigations, evaluate methods and techniques, and suggest possible improvement; • Record observations, measurement, methods and techniques with due regard for precision accuracy and units.

Table 3.2 Analysis of inquiry-based practices in O-Level Chemistry and Physics

O-Level Chemistry	*O-Level Physics*
• Follow a sequence of instructions; • Use techniques, apparatus and materials; • Make and record observations, measurements and estimates; • Interpret and evaluate observations and experimental results; • Plan an investigation, select techniques, apparatus and materials; • Evaluate methods and suggest possible improvements.	• Follow instructions; • Carry out techniques, use apparatus, handle measuring devices and materials effectively and safely; • Make and record observations, measurements and estimates with due regard to precision, accuracy and units; • Interpret, evaluate and report upon observations and experimental data; • Identify problems, plan and carry out investigations, including the selection of techniques, apparatus, measuring devices and materials; • Evaluate methods and suggest possible improvements.

biodiversity and the environment (Figure 3.2) and interpret the relationship between living organisms and their environment (Figure 3.3). The second requires learners to acquire skills in using scientific equipment or follow procedure as depicted by the objectives: *Measure with accuracy and precision (Figure 3.2) and use appropriate technological instruments to collect and analyse data (Figure 3.3)*. The third intends to develop learners' understandings of the scientific approach to inquiry, represented by: *design practical experiments and projects to solve problems (Figure 3.2) and conduct experiments using enquiry (Figure 3.3)*.

The practical work activity specifications and classroom events are well articulated as per the PAAI framework, as evidenced in Figure 3.2 and 3.3. The activity specifications in the two figures show that an investigative approach to practical work is advocated for by the curriculum, which Ramnarain (2011) surmises should commence with learners formulating questions before planning the procedure. Accordingly, Millar's PAAI tool specifies that the first aspect of scientific inquiry should address how a good investigation question can be identified. For instance, in a simple investigation where learners embark on educational tours to bakeries, breweries, dairy industries and sewage works under the topic of microorganisms (Figure 3.3), learners can be engaged in questions such as: Which microorganisms are involved in each of the processes? How do these microorganisms affect the production of yoghurt in the dairy industry? What characteristics do these microorganisms possess which make them suitable for the specific scientific process? Such questions create curiosity in the learners before the visit, which then forms the basis of the investigation.

The Zimbabwean science curriculum endeavours to engage learners in inquiry-based practices. This is also depicted in the Chemistry and Physics syllabi (Tables 3.1 and 3.2) where there are objectives such as: Plan an investigation, select techniques, apparatus and materials (Chemistry) and Identify a problem, design and plan investigations, evaluate methods and techniques, and suggest possible improvement (Physics). According to Millar's (2009) PAAI specification of aspects of scientific inquiry, question identification, planning, selecting equipment and evaluation of methods are key.

Based on the analysis above, the curriculum shows great strides in advancing inquiry-based science teaching and learning. This conclusion about the Zimbabwean science curriculum refutes findings from Mukaro and Stears' (2017) study to determine how teachers interpret the intended curriculum with regard to science process skills in A-Level Biology practical work. Mukaro and Stears pointed out that science process skills are not explicitly stated, but implied within the curriculum, which leaves teachers with the responsibility of identifying the relevant skills in their interpretation of the curriculum documents. The above two researchers also argue that although the curriculum emphasises the teaching of Biology using practical work, it fails to indicate the extent to which practical work should be done. There is an assumption that the teachers can competently interpret the intentions of the curriculum designers. In other words, teachers may fail to interpret the objectives appropriately in order to assist learners in planning investigations.

Teacher enactment of inquiry-based learning in the Zimbabwean science classrooms

The success of the implementation of this inquiry-oriented Zimbabwean science curriculum is dependent on the teachers' willingness, their abilities and the conduciveness of the science classroom context. In concurrence, Ughamadu (2005) propounds that the success of any science curriculum is dependent on the quality of the teachers who implement it. There is a tendency for practical investigations to be equated with traditional laboratory work where learners perform prescribed experiments (normally from a manual or textbook) to reinforce their knowledge of specific scientific concepts. For instance, in a study to determine the nature of A-Level Chemistry teaching and learning in Gweru Secondary schools in Zimbabwe, Mandina (2012) found that 80% of the class time was spent on teacher-dominated and initiated interactions with learners at the expense of engaging learners in inquiry-based practices. It was also found that the expository instruction took up 83% of the class time while only 17% could be regarded as inquiry-based instruction. Expository instruction does not afford learners the opportunity to ask questions, observe, compare, contrast and hypothesise (Cuevas et al., 2005). Because practical work helps to demonstrate concepts learned in theory, from Mandina's study it can be argued that learners fail to grasp concepts in a comprehensible manner as concepts remain abstract. According to Ajaja (2009), such an approach as is prevalent in Zimbabwean schools does not respond to a worldwide call for teachers to plan inquiry-based programmes for learners as well as supporting learners doing inquiry. In such a situation, a scientific discord between philosophical and practical conceptions of inquiry exists (Abd-El-Khalick et al., 2004).

Mandina (2012) attributes the failure by Zimbabwean science teachers to provide inquiry-based instruction to teachers' incompetency, which results from ineffective teacher professional training. Similarly, in a study on issues in teacher supply, training and management in Anglophone Africa, Mulkeen (2010) noted that most teachers have science degrees but they lack teaching qualifications, which compromises how the science subjects are taught. A similar scenario exists in Zimbabwe. These findings concur with the situation in other African countries such as Gambia, Tanzania and Lesotho (Mulkeen, 2010). In their study to examine how Biology is taught at O-Level in Bulawayo Mzilikazi District schools in Zimbabwe, Mwangu and Sibanda (2017) found that teachers expect laboratory technicians to plan, prepare and perform trials before learners are engaged in practical work. It does not enhance learner development of inquiry skills since these technicians do not have the pedagogical skills to assist learners when they are carrying out the investigations. As a result, previous studies on the teaching and learning of practical work in Zimbabwean schools revealed learners' dissatisfaction with the inadequate support and guidance they receive from their teachers during practical work sessions (e.g. Zezekwa, 2006; Mukaro & Stears, 2017).

Accordingly, in his study to determine the impact of practical tests/examinations in developing practical work skills to A-Level Physics learners in Zimbabwe, Zezekwa (2006) noted that most A-Level Physics graduates lack important skills such as setting up experimental apparatus, designing practical work and even recording readings accurately. This shows that although the curriculum prescribes inquiry-based practices in science education, there is a mismatch between the intended curriculum and the implemented curriculum. The Physics curriculum acknowledges the pivotal role that design practical skills play in the learning of the subject. As such, it includes a design component, which develops learners in critical practical knowledge and skills in Physics. Munikwa, Chinamasa and Mukava (2011) investigated teaching strategies used by Physics teachers in seven districts of Zimbabwe in preparing learners for the A-Level Physics practical work. They established that teachers employed traditional strategies (lecture, guided and demonstration experiments), which failed to effectively prepare learners for the design practical examination question. Munikwa et al. (2011) found that most of the teachers (77%) assign only one design practical for the whole term. This further demonstrates that a gap exists between the curriculum stipulations and what happens in reality in Zimbabwean science classrooms.

The following section intends to answer the question: Are assessments aligned with the emphasis given to inquiry-based practices in the curriculum?

Relationship between the prescribed curriculum inquiry-based practices and the examination frameworks

The Zimbabwean science curriculum has provisions at both Ordinary and Advanced Level Biology, Chemistry and Physics for a summative assessment component of practical work.

A-Level Biology, Chemistry and Physics

In Biology, for the two years of studying A-Level, learners are expected to have performed five practical tests, which are recorded and contribute a third of the continuous assessment mark that makes up 30% of the final mark. Summative evaluation of practical work involves a two-and-a-half-hour paper which contributes 14% of 70% towards the final mark. The rest of the 56% comes from three other papers (Multiple choice, Theory-structured and Theory-short free response essay type). Both the continuous and the summative assessments contribute towards the pre-requisite mark entry requirement for university studies, which varies for different science-related degrees and universities. Table 3.3 is an illustration of how inquiry objectives are tested in an A-Level examination. It should be noted that learners' inquiry skills are assessed based on the report that the learner submits at the end of the examination.

O-Level Biology, Chemistry and Physics

Forms 3 to 4 Biology assessment is based on 40% continuous assessment and 60% summative assessment. Continuous assessment of practical work involves administration of two individual or group practical tests per term, which contributes 20% towards the 40% above. Learners are assessed to determine their ability to manipulate apparatus, follow procedures, collect and analyse data, and present and evaluate results. Summative assessment of practical work entails learners sitting for a one-and-a-half-hour paper, which contributes 30% towards the final mark. The remaining 70% comes from the other two papers (Multiple choice and Theory). A similar assessment procedure is followed for both Chemistry and Physics. Based on the analysis above (e.g. Table 3.3), it could be concluded that the assessment tasks are cognisant of the emphasis the science curriculum places on inquiry practices. It is important to explore how the learners perform in practical work examinations.

Extent of learner preparation and performance in practical assessments

The question now is: Can learners pass high school science without having attained the basic practical skills? According to the Zimbabwe Schools Examinations Council (ZIMSEC) reports for the 2010 and 2011 O-Level Biology

Table 3.3 Illustration of how inquiry objectives are being tested in an A-Level examination

Assessment objectives for inquiry as outlined in the curriculum	Examples of experimental skills and investigations from a 2018 Biology specimen examination paper 4 (ZIMSEC, 2018)
1. Follow a sequence of instruction	Prepare a water bath by half filling a 400cm^3 beaker with water and maintain the temperature of the water at about 65 °C. Then …….
2. Use techniques, apparatus and materials	You are required to compare the amounts of reducing sugars in three different fruits.
3. Make and record observations, measurements and estimates	Record the results in a table. Plot a graph of the results in the grid.
4. Interpret and evaluate observations and experimental data	Explain the pattern of the results. Suggest any two improvements that could be made in order to get more accurate results of the sugar content of these fruits.
5. Devise and plan investigations, select techniques, apparatus and materials	You are required to investigate the effect of temperature on the rate of respiration in yeast.
6. Evaluate methods and techniques, and suggest possible improvements	State any three possible sources of error in the investigation. Suggest any two improvements that could be made in order to get more accurate results of the sugar content of these fruits.

The table shows how each inquiry skill is assessed in the final examination using examples of items from the 2018 Biology specimen examination paper.

practical work examinations, learners displayed poor performance as a result of inability to: read the scale on the grid correctly; interpret tabulated data; and draw conclusions from their findings. This poses the pertinent questions on whether learners were well prepared and supervised during practical work and on the presentation of their reports. On that note Ruparanganda, Rwodzi and Mukundu (2013) found that due to poor development of problem-solving skills during laboratory practical work, Zimbabwean Biology O-Level learners perform badly in the final practical work examinations. Examination reports have also consistently indicated that the majority of A-Level Biology candidates lose marks in practical examination not because they do not know the biological concepts that they were tested on, but because they use wrong procedures in designing their experiments (ZIMSEC, 2010–2015), which shows a lack of practice in practical work during the course of study.

The lack of preparation of learners for practical examinations also applies in other science subjects. For instance, in a study to evaluate the impact of current Physics practical assessment methods on skills development at A-Level in Zimbabwe, Zezekwa (2006) questioned the adequacy of the practical skills that high school graduates attain. Zezekwa argues that teachers tend to focus on presentation skills because this is what the examiner assesses and very little attention is afforded to equipment manipulation, investigation and design. Much emphasis is placed on how to write investigation reports at the expense of acquiring inquiry skills. In this context, the researcher questioned: "How can the experimental skills as outlined in the syllabus be fully assessed on the basis of the submitted practical report?" (Zezekwa, 2006, p. 18). Science practical work is better assessed directly by observing learners in action rather than indirectly through written reports (Abrahams, Reiss, & Sharpec, 2013).

Based on the above, important questions to ask are: Do teachers have the knowledge base for implementing inquiry? Does the curriculum create modes of inquiry that teachers can adapt for their different classroom contexts? In order for teachers to effectively enact an inquiry-based pedagogy, they should have a rich base of science content knowledge, and a deep understanding of the learning process. In addition, science teachers should have explicit knowledge of the nature of science and an effective ability to engage learners in investigative practices (Keys & Bryan, 2001). Generally, teachers employ a didactic chalk-and-talk approach in addition to laboratory activities which are mainly of a verification nature. As pointed out by Hahn and Gilmer (2000), a challenge some science teachers face is that they themselves do not have first-hand experience of doing authentic scientific inquiry during their high school and university years. As such, expecting such teachers to properly implement inquiry-based practice in their science classrooms is sometimes beyond their competency.

The next section analyses science teacher professional development in Zimbabwe in order to assess how effectively and competently they are prepared for inquiry-based science teaching.

Teacher preparation and professional development in teaching practical work

In Zimbabwe, pre-service science teacher professional development is offered by three teachers' colleges, where teachers graduate with a Diploma in Education. These teachers are qualified to only teach Combined Science, Biology, Chemistry and Physics up to O-Level. The teachers are then enrolled for a Bachelor of Education at universities for upskilling. Upon qualification, they become eligible to teach A-Level Biology, Chemistry and Physics. Universities also enrol A-Level graduates into the Bachelor of Education Degree to study the different science subjects. Bachelor of Science graduates can also enrol for a Graduate Certificate programme (Mtetwa & Thompson, 2000), so they can become qualified science teachers. The mentioned different forms of science teacher professional development programmes provide teachers with opportunities to perform practical work meaningfully as the universities are well equipped. It can be inferred therefore that the teachers who graduate from these teacher professional development programmes are fully equipped to implement inquiry-based science teaching as espoused in the Zimbabwe science curriculum.

However, there has been increased demand for science teachers in Zimbabwe for the past 10 to 15 years due to deaths caused by HIV/AIDS (UNESCO, 2008) and because of attrition to neighbouring countries. The artificial shortage caused by migration has resulted in Bindura University of Science Education embarking on virtual and distance science teacher development programmes, which extended access to science teacher education in order to meet the high demand (Mpofu et al., 2012). Mpofu et al. (2012) reported limited to no teacher involvement in practical work in these programmes due to shortage of appropriate infrastructure and equipment. Once qualified, such graduate science teachers are therefore compromised in their abilities to implement inquiry-based science teaching in the classroom.

The Zimbabwean government has over the years partnered with other organisations in developing science teachers and infrastructure for effective science teaching and learning. One such innovation was the Zimbabwe Science project, which provided science kits with basic science equipment to rural schools without any science infrastructure. The project won the UNESCO Jon Amos Commenius award for excellent innovation in teaching science. Another example that was aimed at in-service training science teachers was the Quality Education in Science Teaching (QUEST) project that decentralised in-service training of science teachers using the cascade model. Further to this, the Science Education In-service Teacher Training (SEITT) programme focused on staff development of advanced science and mathematics teachers with the help of more qualified and skilled resource teachers. This project established resource centres equipped with laboratory stuff (funded by the Netherlands government). Better Environmental Science Teaching (BEST) catered for primary science education with a focus on training teachers to use the environment as a laboratory for science teaching and learning (funded by the German government).

Continuing science teacher professional development is largely based on the cascade method that is based on teacher-cluster groups, provincial professional learning centres, and resource persons who can be experienced science teachers (Mtetwa & Thompson, 2000). Through in-service training workshops, science teachers are professionally developed in new teaching approaches such as inquiry-based approaches through engaging in practical work in resource centres. Normally resource teachers (expert teachers) are well qualified, and in most cases have a formal in-service qualification such as a Post Graduate Diploma in Science Education. This qualification is offered by the University of Zimbabwe in conjunction with the Netherlands government as funders. The description above shows that science teachers have different avenues for professional development. The following section explores the extent of technology use in teaching inquiry-based practices.

Technology use in the teaching of inquiry-based practices

Given the topical buzz-phrase 'the Fourth Industrial Revolution', science teaching is also keeping abreast with these technological changes. A review of pertinent literature on the use of technology in Zimbabwean science classrooms suggests that the use of animations, simulations and virtual laboratories is sparse. This is a concern, as such technologies can provide opportunities for learners to actively engage in practical activities (Tüysüz, 2010), and also help to address the problem of ill-equipped science laboratories. Mwangu and Sibanda (2017) found in their study that only a few teachers and schools have access to e-learning software or virtual laboratories in Zimbabwe, and hence most of them utilise traditional textbooks, charts, posters and models to teach Biology. However, findings from studies in other parts of the world have shown that PowerPoint is the most widely used ICT-based instructional technology, while animations were the least used ICT-based instructional technologies (Maharaj-Sharma, Sharma & Sharma, 2017). Computer-aided simulations, smartboards and virtual labs are other ICT-based instructional technologies used by teachers in other countries. In Zimbabwean schools, contextual factors make it difficult for public school teachers to use such ICT-based instructional technology. The country is plunged in serious load shedding and very few non-governmental schools can afford generators, computers, iPads and other ICT tools for their learners. It is therefore safe to infer that very few teachers are using ICT-based instructional technology to implement inquiry-based science teaching and learning in their schools in Zimbabwe. The next section explores possible challenges and problems in practical work implementation.

Possible challenges and problems in the implementation of inquiry-based practices

In the real science classrooms, the implementation of practical work is sometimes compromised because of various factors. In their study exploring the possibilities of using a project approach in place of regular practical work in Biology, Ruparanganda et al. (2013) explained the dire situation of the lack of or poorly equipped laboratories in most rural secondary schools in Zimbabwe. Mandina (2012) also found that A-Level Chemistry instruction fell below the required standards due to shortage of equipment and consumables. She also noted that A-Level Chemistry learners in non-government schools (more resourced) were more exposed to practical work than their counterparts in government schools (poorly resourced). The problem is compounded by a lack of skilled support staff in the form of laboratory technicians/assistants. Because of large numbers and poorly resourced laboratories, teachers resort to using demonstrations (Mwangu & Sibanda, 2017), which do not prepare learners adequately to undertake individual practical examinations at the end of the academic year. As such, most learners have continued to perform badly in science practical examinations compared to theory papers (ZIMSEC Examiners'

Reports, 2014). Accordingly, Mwangu and Sibanda (2017) attribute teachers' inadequate knowledge and skills in teaching inquiry skills as the major cause of learners' poor performance.

Chirikure, Hobden and Hobden (2018) conducted a study on A-Level Chemistry learners' approaches to investigations from a learning perspective in the Zimbabwean educational context. Approaches to investigations refer to the manner in which learners plan and carry out investigations, which can result in deep, strategic or surface learning (Chirikure et al., 2018). The approaches are inextricably linked to the quality of learning and performance in these practical activities. The learners were found to be predominantly strategic in their approach to investigations. Because in this approach learners are wary about achieving excellent grades, their focus was on maximising examination preparation following previous assessments (Donnison & Penn-Edwards, 2012). Learners tended to tailor their answers according to examiners' requirements. Their approaches were influenced by contextual factors such as instructional strategies and prior Chemistry practical experience, which tend to be dominated by the end product of success in examinations. The findings of this study provide useful insights for curriculum designers on the enactment of the practical component in Chemistry and all sciences with respect to investigations. A shift from the traditional high stakes final examination to a school-based continuous assessment of investigations might be a viable move towards a deep approach to investigations and a greater emphasis on developing process skills.

In a technical report on a diagnostic study on the status of STEM in Zimbabwe, Gadzirayi, Bongo and Bhukuvhani (2016) confirmed the shortage of laboratories, equipment, chemicals, and other 'paraphernalia' required in STEM education in the country. Mandina (2012) also found that most schools have inadequate science budgets and hence lack funds to purchase enough equipment and laboratory apparatus as well as consumables in the form of chemicals. This in itself acts as a barrier to the successful implementation of the A-Level science curriculum. In several studies, teachers also mentioned that there is insufficient time to prepare and implement inquiry practices in their classrooms due to the overloaded school curriculum (Zezekwa, 2006; Mandina, 2012; Ruparanganda et al., 2013).

In another study, Vhurumuku et al. (2006) investigated the proximal and distal images of the nature of science that A-Level learners develop from their participation in Chemistry laboratory work. The study found that very rarely do the teachers use textbooks as sources of problems for their learners' laboratory work. The laboratory instruction was almost without exception examination-oriented. The level of inquiry was generally low, with most of the experiments being verificationistic and illustrative. Both the teachers and the learners placed great value in ensuring adequate preparation for the laboratory work examination at the expense of developing inquiry skills. There is a need for teachers to receive continuous support and guidance for the successful implementation of inquiry-based science teaching (Abd-El-Khalick et al., 2004).

Conclusion

In assessing the advances in inquiry-based science education in Zimbabwean high schools, this chapter has analysed both O-Level and A-Level Biology, Chemistry and Physics syllabi. The extent to which inquiry-based practices are implemented and portrayed in the science classrooms and assessment tasks was also explored. In policy, the Zimbabwean science curriculum endeavours to engage learners in inquiry-based practices. This is articulated in the curriculum objectives, practical work activities and the events planned for the classrooms. The analysis reveals that learners' appreciation of inquiry-based learning is an explicit goal of the curriculum. To a certain extent there is an assumption that the teachers can competently interpret the intentions of the curriculum designers; however, this assumption is unfounded in most cases. In terms of implementation of the inquiry-based practices in the classroom, it can be concluded that its success is dependent on the teachers' willingness, their abilities and the conduciveness of the science classroom context. While there is a myriad of contextual factors hampering inquiry-based teaching and learning in Zimbabwe, the most disturbing reality is the teachers' failure to provide inquiry-based instruction to learners due to incompetency and negative attitudes in that regard. As such, in certain instances learners performed badly in the practical examinations due to a lack of preparation for practical examinations.

Implications and recommendations

The chapter findings and chapter conclusion show that the level of inquiry-based science teaching and learning is below expectations, which hampers both the economic and technological development of the country. There is a need for continued teacher professional development to develop teachers' pedagogical knowledge and strategies, which will enhance learners' competencies in practical work. This chapter argues that the current science teachers should be retrained and in-serviced to give them a better orientation on what is expected of them, considering that some of them were trained before the dawn of technology. Such refresher courses are likely to expose science teachers to current methods of teaching, particularly the use of inquiry-based practices. The Ministry of Primary and Secondary Education as it is known in Zimbabwe should provide the necessary infrastructure and enabling environment to make science education thrive through inquiry-based teaching and learning. The different schools should revive the cluster-based training and resource centres, which made great strides in the past in equipping science teachers with knowledge and skills for utilising and sharing local resources and expertise, rather than waiting for government initiatives.

The many Zimbabwean universities mandated to train pre-service science teachers should benchmark their science teacher education curricula with other universities in the continent and across the world. This will enhance the quality of teachers they produce in terms of current pedagogies, especially in the current Fourth Industrial Revolution era where technology has taken over.

References

Abd-El-Khalick, F., Boujaoude, S., Duschul, R., Lederman, N. G., Hofstein, R., Hofstein, R., Niaz, M., Treagust, D., & Tuan, H. (2004). Inquiry in science education: International perspectives. *Science Education*, 88(3), 398–419. doi:10.1002/sce.10118

Abrahams, I., Reiss, M. J., & Sharpec, R. M. (2013). The assessment of practical work in school science. *Studies in Science Education*, 49(2), 209–251.

Abrams, E., Southerland, S. A., & Evans, C. A. (2007). Inquiry in the classroom: Necessary components of a useful definition. In E. Abrams, S. A. Southerland, & P. Silva (Eds.), *Inquiry in the science classroom: Realities and opportunities* (pp. 1–45). Greenwich, CT: Information Age Publishing.

Ajaja, O. P. (2009). Evaluation of science teaching in secondary schools in Delta State 2: Teaching of the sciences. *International Journal of Educational Science*, 1(2), 119–129.

Blanchard, M. R., Southerland, S. A., Osborne, J. W., Sampson, V. D., Annetta, L. A., & Granger, E. M. (2010). Is inquiry possible in light of accountability? A quantitative comparison of the relative effectiveness of guided inquiry and verification laboratory instruction. *Science Education*, 94(4), 577–616. doi:10.1002/sce.20390

Chirikure, T., Hobden, P., & Hobden, S. (2018) Exploring Zimbabwean Advanced Level Chemistry students' approaches to investigations from a learning perspective, *African Journal of Research in Mathematics, Science and Technology Education*, 22(1), 60–69.

Crawford, B. A. (2012). Moving the essence of science into the classroom: Engaging teachers and students in authentic science. In K. C. Tan, & M. Kim (Eds.), *Issues and challenges in science education: Moving forward* (pp. 25–42). Dordrecht: Kluwer.

Cuevas, P., Lee, O., Hart, J. & Deaktor, R. (2005). Improving science inquiry with elementary students of diverse backgrounds. *Journal of Research in Science Teaching*, 42, 337–357.

Donnison, S., & Penn-Edwards, S. (2012). Focusing on first year assessment: Surface or deep approaches to learning? *The International Journal of the First Year in Higher Education*, 3(2), 9–20.

Edelson, D. C. (1998). Realising authentic science learning through the adaptation of scientific practice. In B. J. Fraser & K. G. Tobin (Eds.), *International handbook of science education* (Vol. 1, pp. 317–331). Dordrecht: Kluwer.

Gadzirayi, C. T., Bongo, P. P., & Bhukuvhani, C. (2016). *Diagnostic study on status of STEM in Zimbabwe*. Technical report. Bindura: Bindura University of Science Education.

Gott, R., & Duggan, S. (2002). Problems with the assessment of performance in practical science: Which way now? *Cambridge Journal of Education*, 32(2), 183–201.

Hahn, L. L., & Gilmer, P. J. (2000, October). *Transforming pre-service teacher education programs with science research experiences for prospective science teachers*. Paper presented at the annual meeting of the South-eastern Association for the Education of Teachers in Science, Auburn, AL.

Hodson, D. (1996). Practical work in school science: Exploring some directions for change. *International Journal of Science Education*, 18(7), 755–760.

Hofstein, A., & Lunetta, V. N. (2004). The laboratory in science education: Foundation for the 21st century. *Science Education*, 88(1), 28–54.

Keys, C. W., & Bryan, L. A. (2001). Co-constructing inquiry-based science with teachers: Essential research for lasting reform. *Journal of Research in Science Teaching*, 38(6), 631–645.

Maharaj-Sharma, R., Sharma, A., & Sharma, A. (2017). Using ICT-based instructional technologies to teach science: Perspectives from teachers in Trinidad and Tobago. *Australian Journal of Teacher Education*, 42(10), 22–35.

Mandina, S. (2012). An evaluation of advanced level chemistry teaching in Gweru district schools, Zimbabwe. *Asian Social Science*, 8(10), 151–159. doi:10.5539/ass.v8n10p151

Millar, R. (2004). *The role of practical work in the teaching and learning of science*. Washington, DC: National Academy of Sciences.

Millar, R. (2009). *Analysing practical activities to assess and improve effectiveness: The Practical Activity Analysis Inventory (PAAI)*. York: University of York.

Millar, R., Tiberghien, A., & Le Maréchal, J. F. (2002). Varieties of labwork: A way of profiling labwork tasks. In D. Psillos & H. Niedderer (Eds.), *Teaching and learning in the science laboratory* (pp. 9–20). Dordrecht: Kluwer Academic.

Ministry of Primary and Secondary Education, (2015). *Biology syllabus Forms 3–4 (2015–2022)*. Harare: Curriculum Development Unit and Technical Services.

Mpofu, V., Samukange, T., Kusure, L. M., Zinyandu, T. M., Denhere, C., Huggins, N., Wiseman, C., Ndlovu, S., Chiveya, R., Matavire, M., Mukavhi, L., Gwizangwe, I., Magombe, E., Magomelo, M., & Sithole, F. (2012). Challenges of virtual and open distance science teacher education in Zimbabwe. *The International Review of Research in Open and Distance Learning*, 13(1), 208–219.

Mtetwa, D. K. J., & Thompson, J. (2000) Towards decentralised and more school-focused teacher preparation and professional development in Zimbabwe: The role of mentoring. *Journal of In-Service Education*, 26(2), 311–328. https://doi.org/10.1080/13674580000200119

Mukaro, J. P., & Stears, M. (2017). Exploring the alignment of the intended and implemented curriculum through teachers' interpretation: A case study of A-level biology practical work. *Eurasia Journal of Mathematics, Science & Technology Education*, 13(3),723–740.

Mulkeen, A. (2010). *Teachers in Anglophone Africa: Issues in teacher supply, training and management*. New York: The World Bank.

Munikwa, S., Chinamasa. E., & Mukava, M. 2011. A study of teaching Advanced level Physics practical and solution approach to practical questions. *Journal of Innovative Research in Education*, 1(1), 36–48.

Mwangu, E. C., & Sibanda, L. (2017). Teaching Biology practical lessons in secondary schools: A case study of five Mzilikazi District secondary schools in Bulawayo Metropolitan Province, Zimbabwe. *Academic Journal of Interdisciplinary Studies*, 6(3), 47–55.

National Research Council (NRC). (1996). *National science education standards*. Washington, DC: National Academic Press.

Ramnarain, U. (2011). Teachers' use of questioning in supporting learners doing science investigations. *South African Journal of Education*, *31*, 91–101https://doi.org/10.15700/saje.v31n1a410

Ruparanganda, F., Rwodzi, M., & Mukundu, C. K. (2013). Project approach as an alternative to regular laboratory practical work in the teaching and learning of biology in rural secondary schools in Zimbabwe. *International Journal of Education and Information Studies*, 3(1), 13–20.

The United Nations Educational, Scientific and Cultural Organization (UNESCO) (2008). *World challenges*. Operational guidelines for the implementation of the World Heritage Convention. Paris: UNESCO World Heritage Centre.

Tüysüz, C. (2010). The effect of the virtual laboratory on students' achievement and attitude in chemistry. *International Online Journal of Educational Sciences*, 2, 37–53.

Ughamadu, K. A. (2005). *Curriculum: Concept, development and implementation*. Onisha: Emba Printing and Publishing Company Ltd.

Vhurumuku, E., Holtman, L., Mikalsen, O., & Kolsto, S. D. (2006). An investigation of Zimbabwe high school chemistry students' laboratory work-based images of the nature of science. *Journal of Research in Science Teaching*, 43(2), 127–149.

Zezekwa, N. (2006). The impact of practical tests in developing practical work skills to A-level physics students. *Southern African Journal of Education, Science and Technology*, 1 (1), 17–22.

Zimbabwe Schools Examination Council (ZIMSEC) (2010–2015). *Biology examination reports*. Harare: Government Printers.

Zimbabwe Schools Examination Council (ZIMSEC) (2018). *General Certificate of Education Advanced Level Biology 6030/4 specimen paper*. Harare: Government Printers.

Chapter 4

The role of practical work in the teaching of science in Nigerian schools

Johnson Enero Upahi and Oloyede Solomon Oyelekan

Introduction

Practical work plays a prominent role in science and science education. It has been described as a distinctive feature of science education (Millar, 2004). Practical work provides learners with opportunities to engage in and have first-hand experience of what it means to carry out scientific investigations (Hofstein, Kipnis & Abrahams, 2013). According to Millar (2004), *practical work* is a term that is commonly used to refer to science teaching and learning activities in which students (either individually or in small groups) are involved in manipulating and/or observing real objects and materials (e.g. determining which selection of objects are magnetic, carrying out and observing flame tests). Practical work has generally been referred to as hands-on activities or doing science, which may or may not take place in the laboratory (Millar, 2011). In this sense, practical work includes a broad category of activities that include cookbook or recipe tasks, experiments, investigations, discovery learning tasks, and science inquiry. However, school science teachers and students have been reported to use the words practical, experiment, practical work and scientific investigation interchangeably in describing activities that characterize practical work, and this seems to create confusion about what practical work really means. A more helpful discussion can be found in Ramnarain and Kibirige (2010) where the authors explain the differences and relationships among practical work, investigations and inquiry.

Wellington (1998) argues that when these activities are characterized on the basis of location (i.e., where the activities take place) but not on the nature of the activities undertaken, it would seem more appropriate to rather consider them as *laboratory work* than as practical work. This position is informed by a view that most of the laboratory work carried out in school science, around the world, including Nigeria, takes place in what White (1988) described as "purpose-built laboratories" (p. 8). However, since location is not a critical feature in characterizing the broad category of activities mentioned earlier, these activities can also be thought of as practical work.

Apart from the broad focus of this book, our preference for the term *practical work* emanates from an understanding that observation or manipulation of objects could extend beyond the school laboratory such as having to study an aspect of biology in the field/out-of-classroom setting. To put it succinctly, practical work broadly includes experiments, practical activities and scientific investigations and laboratory work. In our subsequent discussion, we bear in mind that activities which characterize or define practical work may not necessarily take place in a science laboratory (Millar, 2011).

The purpose of this chapter is to examine the role and purpose of practical work in the teaching of science in Nigeria. First, it is important to note that our intention is not to condemn the current practice of practical work but to critique it and contribute to the debate on practical work from an African perspective, and reposition it for the future of science education within the continent and globally. Second, it is equally our desire to use the platform of this book to inform teaching practitioners, researchers, curriculum developers and policy makers within our context of the need to re-examine practical work with the sole aim of improving science teaching and learning. Going forward, we begin our conversation/discussion on the role of practical work with an overview of the goals of science and science education.

The goals of science and science education

An understanding of the goals of science and science education is an important point of departure to begin conversations/discussions on the role of practical work in science and science learning. Millar (2004) summarizes the goals/aims of science learning as:

1   helping students gain an understanding of much of the established body of scientific knowledge that will be appropriate and relevant to their needs, interests and capacities; and
2   developing students' understanding of the methods by which scientific knowledge has been established, and the justifications for knowledge claims about the nature of science.

These goals of science education have been observed to be conflated with the goals of science in literature (Moeed & Anderson, 2018; Osborne, 2014). This appears to have generated some confusion surrounding the role of practical work in science and science learning. While some science educators hold the view that students can acquire scientific knowledge through their own designed scientific investigations/inquiry – a view that Millar (2004) believes erroneously draws a parallel between the purpose of learners and scientists, others have the orientation that practical work in the school science curriculum serves the purpose of helping learners understand the established body of scientific knowledge. A view consistent with the latter was expressed by Osborne

(2014) who makes a clear distinction between the goals of science and science learning. He argues that the goal of science learning is to help students learn about the established body of knowledge in science that modern culture has built about the material world that surrounds us, while the goal of science is to create new knowledge and critique existing theories in the light of new evidence.

In a similar perspective, Moeed and Anderson (2018) caution that the rigorous procedures scientists develop and follow in investigating science disciplines to generate valid and reliable evidence should not be conflated with the procedure students follow to plan and carry out investigations during practical work. While students may design experiments, collect data, interpret and generate explanations from evidence, this does not mean that such activities will create new knowledge. Rather, the activities mimic the procedure scientists use, with students solving problems and answering questions about established knowledge that relates to the material world.

Following these submissions, we affirm that the goal of science learning is really not to discover new knowledge but to help learners develop an understanding of existing, consensually agreed and well-established knowledge. This includes developing substantial understanding of the models that scientists use to explain the behaviour of the material world, their methods of inquiry and rationale for knowledge claims.

In summary, the definition of practical work presented in the introduction section describes it as an activity that is limited to a school laboratory. While the activities of scientists and those required for science learning are also not limited to a school laboratory, it is important to understand that the role of practical work in the *enterprise* of science arguably differs from that of science learning. Scientists engage in practical work to discover new knowledge, but science learning uses practical work to engage learners in scientific investigations of phenomena whose interpretations are well-established knowledge (Millar, 2004). In the next section we trace how science evolved within the school curriculum in Nigeria. Since school science is now firmly rooted in the laboratory, the subsequent section examines practical work or laboratory activities as an essential item in the school science curriculum.

Science in the Nigerian school curriculum

The history and development of science education in Nigeria has its roots in the colonial era. Science was first introduced as part of general education by the Christian missionaries in mission schools between 1859 and 1920. It became widely accepted in government-controlled schools after the report of the African Education Commission sponsored by The Phelps–Stokes Fund of America. The justification for the inclusion of science in the school curriculum came about when the Commission observed that:

[T]he great achievements of modern times are largely in the realm of the physical sciences. Physics, chemistry and biology have revolutionized many of the industrial and social activities of mankind. No phase of secondary education is more vital than the instruction of the pupils in the elements of these sciences. It is of the utmost importance that the pupils should gain power to apply the facts and principles of science and to interpret natural phenomena.

(Jones, 1922, p. 8)

According to Jegede (1988), analysis of the science curricula constructed for school science before 1960 showed that they mainly focused on scientific facts and principles with little emphasis on the application of the scientific knowledge. By this means, science teaching probably followed suit and students may have ended up learning science content without sufficient understanding of its relevance to their daily living and decision-making processes (Abimbola, 2009).

The major science education curriculum reform movements began from the uproar that occurred in the United States (US) and United Kingdom (UK) after the Soviet Union launched its first space satellite, Sputnik 1, on 4 October 1957. This informed the science curriculum reform initiatives of Biological Sciences Curriculum Study (BSCS), Chemical Education Material Study (Chem Study or CHEMS), Physical Science Study Committee (PSSC), Intermediate Science Curriculum Study (ISCS), Introductory Physical Science (IPS), Earth Science Curriculum Project (ESCP) in the US; and Nuffield Combined Science (NCS), Nuffield Secondary Science (NSS), School Council Integrated Science Project (SCISP), among others in the UK.

In Nigeria, science education reforms took a cue and closely followed the reforms in the US and UK. The reforms gave rise to several programmes that include: The African Primary Science Programme (APSP), Basic Science for Nigerian Secondary Schools (BSNSS), The Ife Yoruba Six-Year Primary Project (YSPP), The Midwest (formerly, Bendel State) Primary Science Project (MPSP), Primary Education Improvement Project (PEIP), Nigerian Secondary Schools Science Project (NSSSP) and the Nigerian Integrated Science Project (NISP). The NISP was a reform initiative pioneered by the Science Teachers' Association of Nigeria (STAN), with financial support from the Ford Foundation through the Comparative Education Study and Adaptation Centre (CESAC) and the Centre for Curriculum Development Overseas (CREDO). Two integrated science courses were produced for use in the first two years of secondary education for the first time.

These reforms were coordinated by a science curriculum development committee that drew participants from STAN, CESAC, the West African Examinations Council (WAEC), the Ministries of Education and the British Council. This committee revised and updated existing West African School Certificate syllabi and produced alternative ones in physics, chemistry and biology. The alternative syllabi were introduced into some secondary schools in

1974 and still co-exist with the revised syllabi of the West African Examination Council (Iyiola & Wilkinson, 1984).

Jegede (1988) noted that while these earlier programmes departed significantly from the expository to a more progressive method of science teaching of what science is and how the scientist works, only minimal changes have been noticed in the manner in which learners learn in science classrooms. He further observed that science teaching is dominated by a focus on the facts and principles of science. He added that teachers were neither positively disposed towards the changes, nor effectively trained to cope with the demands of the new curricula.

The current science curricula used in schools were adapted and revised from the 1985 editions developed by CESAC. The science curricula were written by university lecturers and senior school science teachers and published in 2007 by the Nigerian Educational Research and Development Council (NERDC) and adopted by the Federal Ministry of Education for use in schools in 2009 (NERDC, 2009).

In planning the revised science curricula, a spiral approach to sequencing a science course was adopted. The spiral approach allows concepts/topics to be taught in greater depth as the course progresses, in such a way that a particular concept can run throughout the three years of senior school education. In implementing the science curriculum, curriculum developers affirm that the content should be designed with a focus on practical activity that will take advantage of locally available materials within the learners' context. It would seem logical to assume that the successful implementation of science curricula partly relies on practical work. How then is practical work reflected in the science curriculum? This question forms our discussion in the next section.

Practical work in the school science curriculum

The notion that practical work is an essential part of the school curriculum is a view that is shared by many. Practical work has continued to be a standing item in several curriculum statements, with valid claims to justify its inclusion (Needham, 2014). In the UK, for instance, the written evidence submitted by the Department forEducation to the House of Lords Science and Technology Committee, in canvassing for the place of practical work in the school science curriculum states that:

> [P]ractical science delivered with flair and knowledge can help pupils understand scientific concepts and ignite their interest in physics, chemistry and biology. Practical science is also an important part of scientific knowledge and teaches pupils about the empirical basis of scientific enquiry.
>
> (Department forEducation, 2011, p. 63)

In Nigeria, a similar view is expressed in the senior secondary education science curricula for SS 1–3. Senior school one to three (SS 1–3) represents a senior school level of education in Nigeria, which is equivalent to Grade 10–12 in countries that use the grade system. For instance, in the chemistry curriculum, one of the objectives of the curriculum clearly states that "the curriculum content focuses on practical activity with emphasis on locally available materials. This is to imbue the learners with the spirit of enquiry... will enable the learner to achieve his/her maximum potential in the subject of chemistry and its various applications" (NERDC, 2009, p. 5). The NERDC develop curriculum for the different science subjects such as physics, chemistry and biology. These curricula provide the frame where the states' ministries of education/schools prepare their own subject-specific curriculum. The anticipation that practical work will drive the implementation of curriculum content for students' learning is commendable but the fact that practical work does not seem to have helped students to learn in the way science educators anticipated remains an issue of concern to the science education community (Abrahams, 2011; Millar, 1999, 2004; Millar & Abrahams, 2009; Wellington, 1998).

In the chemistry curriculum document, it is repeatedly stated that teachers will guide students, for instance, to "perform experiments to determine the solubility of substances...carry out experiments on the removal of hardness by boiling and addition of washing soda..., prepare the solution of common states" (NERDC, 2009, pp. 17–19). These extracts further illustrate the prominent place practical work holds in the teaching and learning of the subject of chemistry in the school curriculum. However, these practical tasks appear to be teacher-led activities and share the characteristics of structured inquiry. While the students may develop some basic skills of observing, classifying, measuring and recording, it is not certain whether they will be able to learn the science ideas behind such "structured" investigations as students do not have the opportunity to perform their own investigations where they can gather data, look for evidence, critique the evidence and make informed decisions about the investigations.

To address this drawback for which science educators have contested the conceptual understanding of practical work in school science, one approach that has been found and tested to stimulate students' thinking and discussion is the Predict-Observe-Explain (POE) strategy and its other variants (White & Gunstone, 1992). With the POE strategy, students first predict what they would expect to happen in a practical task; they are then required to write them down, carry out the tasks, make observations and finally explain what they have observed to see whether what they predicted eventually occurred. But unfortunately, this approach is not frequently adopted by teachers in scaffolding students' thinking about the science ideas behind practical activities. The challenge is either that the teachers do not have sufficient subject knowledge on how to use the POE during practical tasks, or they are constrained to design practical work that is primarily driven by the demands of the national assessment policy framework. With these constraints, the POE approach rarely prevails during practical tasks in school science. However, when the POE approach is used in practical work and students' ideas have been identified at the level of making predictions, this could create room to challenge their misconceptions/alternative conceptions, and to also accept or refute their claims when investigating scientific phenomena (Millar, 2004). More detailed evidence of students' learning from practical work in school science is the focus of the next section.

Evidence of students' learning from practical work in school science

In Nigeria, there is limited research on students' learning gains from practical work or hands-on activities in laboratory settings. While there is ample research on practical work in school science, comparatively few international studies report that students learn much when they engage in hands-on activities (Anderson, 2012; Hume & Coll, 2008; Millar, 2010; Moeed, 2010). Millar (2010) draws on the work of Jean Piaget to advance justification for practical work in the laboratory setting. Millar maintains that through practical experience of observing science phenomena, students can generate sensory data that can be *assimilated* into the existing schemas or changed to *accommodate* the new data/information in order to re-establish *equilibrium* between internal and external realities. It is through observations of phenomena and manipulations of objects and materials that students can construct their own representations of the natural world and are more likely to be able to establish links between domains of objects and ideas (Millar, 2004).

Some other science educators have criticized the role of practical work and argue that it contributes little to science learning (e.g. Hodson, 1991; Millar & Abrahams, 2009; Osborne, 1998). According to Millar (2010), the basis of such criticism hinges on the fact that unlike the domains of knowledge, the domain of ideas really requires practical work and when an experiment is designed with intended learning objectives that learners/students will be able to learn a concept/relationship/theory/model within a one-shot practical task, Millar expressed doubts as to whether students can grasp a new scientific concept or theory in one practical session. In the light of these explanations, the current practice of practical work in the Nigerian context does not seem to be promoting effective science learning, especially in the domain of abstract ideas that requires a gradual process over a period of time for the production of conceptual understanding.

The prevalent report in the literature on the perceived gains of practical work relate to motivational benefit (e.g. enjoyment and interest), positive attitude towards science and the development of skills and scientific knowledge (Hodson, 1992). In New Zealand for instance, Moeed (2010) reports that students see investigative work as an enjoyable alternative to written work. In the UK, Toplis (2012) also report students' views on science learning through practical work: students view it as interesting and activity-based, with autonomy and opportunity to engage with their peers in practical tasks. The students noted that practical work gives them some measure of control over their learning and it is preferred to writing and rote learning. While there is euphoria that practical work is motivating and exciting, and stimulates interest and enthusiasm, there is little to suggest that it contributes to the epistemological learning of science concepts. When students are asked about their learning gains in science investigations, their responses largely focus on the processes they followed in carrying out investigations rather than the purpose of the science investigation or how much procedural knowledge they have developed, and the nature of science investigation they understood. The effectiveness of practical work for science learning has therefore been questioned (Moeed, 2010). Students' learning gains in practical work must transcend beyond being fun to include experiences of their natural world and an understanding of how knowledge is constructed (Millar, 2004). In this regard, the POE approach does hold some promise and should be more readily exploited in Nigerian science classrooms.

The nature of assessment of practical work

Practical tasks in school science take the form of hands-on activities where students are required to handle and manipulate objects and materials. In the Nigerian context, a variety of practical tasks are included in school science (physics, chemistry and biology) spanning the three years of senior school education. In carrying out these investigations, students are assessed on their ability to follow instructions, handle materials, make observations, manipulate and use apparatus effectively, measure accurately, gather data and record appropriately. However, this type of assessment underpinning practical tasks that measures science process skills is not formative. It is only formative if the results or feedback are used by the teachers to improve students' science learning.

Since Nigeria operates an education system that is driven by high-stake assessment, the West African Examinations Council (WAEC) is one of the foremost examination bodies that conduct the West African Senior School Certificate Examination (WASSCE) as a summative assessment to mark the end of senior secondary education. It is a regional examination conducted every year in five countries, namely: Nigeria, Ghana, Gambia, Sierra Leone and Liberia. The WASSCE is a qualitative and reliable examination in West Africa that purportedly has a strong influence on learning, teaching and assessment (Ojerinde, 2011).

WAEC registers school and private candidates for May/June and November/December WASSCE, respectively. The May/June WASSCE is an examination offered in Summer, and it is for school candidates who are completing their senior school education. Candidates who sit this examination wear distinctive uniforms as prescribed in the standards established by the schools' boards. By August of every year, the results are released. For the November/December

WASSCE, formerly known as General Certificate Examination (GCE), the examination is offered in Autumn by school and private candidates who may have outstanding subjects to correct before seeking admission into tertiary institutions.

For WASSCE in physics, chemistry and biology, there are three papers – Paper 1, Paper 2 and Paper 3. Candidates are usually required to take either Papers 1 and 2 or Papers 2 and 3 only. Paper 1 is a two-hour practical test in the May/June WASSCE, while Paper 3 also tests practical knowledge but is an alternative to Paper 1 for the November/December WASSCE. Each paper is comprised of three questions, and candidates are required to answer all the three questions. Two of the questions are on quantitative (acid-base titrations) and qualitative (flame tests, iodine tests etc.) analyses, while the third question tests candidates' familiarity with the practical tasks suggested in each of the teaching syllabi. The questions constitute 25% of the total marks (West African Examinations Council [WAEC], 2005).

Paper 2 is a three-hour theory paper that covers the entire syllabus and carries about 150 (75%) marks of the total marks of the external examination. The paper has two parts: Part A and Part B. Part A contains 50 multiple-choice questions, which candidates are required to answer within 60 minutes for 50 marks. Part B comprises three sections, namely: Sections I, II and III. While there are usually four essay questions in Section I out of which candidates are expected to answer only three, candidates are also required to answer only one question from either Section II or III. Each of the four questions is allocated 25 marks.

The nature of these practical tasks is such that it is mainly driven by summative assessments. The question of whether this practice does not limit the goal/purpose of doing science investigations remains a critical issue. According to Donnelly, Buchan, Jenkins and Welford (1993), the outcomes of such activities may not really be used for formative assessments to improve students' learning of science. These descriptions clearly reflect how students' learning are assessed in practical work at the end of their senior school education. What then is the role of practical work in school science? This important question is the focus of the next section.

The roles of practical work in the teaching and learning of science in Nigerian schools

The Nigerian geographical environment provides ample opportunities for scientific investigations. The weather is friendly and the climate is diverse. Hence, this provides teachers and students the opportunity to explore the environment for practical tasks. The extent to which this can be achieved is limited by logistics, finance and availability of laboratory equipment/facilities.

Practical science work cuts across the disciplines of science. Sometimes you start in the laboratory and end up on the field; sometimes it is vice-versa. The roles that practical science work could play in the science learning of students are directly related to the facilities available for the students to learn science among other factors. A student who has been to a pond with his teacher to collect Spirogyra and examined it under a microscope in the laboratory has a different learning experience from a student who is taught about Spirogyra using diagrams alone. Similarly, the experience of students who are taken through a computer-simulated chemistry experiment will be different from that of students who actually carry out the experiment on their own.

School science practical work is confronted with a myriad of challenges, some of which will be discussed in the next section. However, it is important to note that these challenges affect school science to varying degrees. With or without these challenges, science practical work when conducted has significant roles to play in the teaching and learning of science. Some of these roles are explained as follows:

Taking science from the realm of abstraction to the realm of reality. The abstraction that characterizes the explanation of some scientific theories and principles are made real during science practical lessons. For example, stoichiometric principles as expressed in chemical equations are brought to reality when quantitative measurements are practically determined between reactants and products in a chemical reaction. This is what validates theoretical explanations that could not be physically observed. The relationship among mole, mass and molar mass as is explained in chemical equations and mathematical calculations can only be made real when practical experiments are conducted, and the quantitative measurements agree with the mathematical calculations emanating from the equation of the reaction. Similarly, in biology, when students are told in class that microbes are in the air, under their nails and in their mouth, they cannot see them. However, with practical work designed to prove these scientific facts, science is taken from the level of 'unseen' to the level of 'seen'. This important role of science practical work is to provide students with first-hand experience of phenomena.

Source of motivation. Scientific experiments can be very interesting and fascinating. The practical experience of some scientific experiments has the potential to stimulate students' interest to like science, and consequently, spur them to pursue a career in science and technology. When a student is well-motivated by hands-on activities, s/he can put in additional effort to pursue a science-related career. The first-hand practical experience that hands-on activities affords students when engaging directly with materials can be a source of motivation to them. This is an entirely different scenario to students merely sitting in the class either to listen to a teacher demonstrate or explain some scientific theories and principles to them. One sure way of sustaining students' interest is to make science practical-oriented.

Psychomotor skills. The active involvement of students in science practical enables the use of their hands to manipulate objects or materials. Hence, they acquire manipulative skills which are of vital importance in the scientific process. Their ability to utilize scientific equipment efficiently is enhanced and they become familiar with the use of laboratory materials by their constant use. Psychomotor skills are of importance in science as they facilitate a high level of precision and accuracy in measurements. In science teaching, psychomotor skills are fundamental skills that teachers need to build in their students. They can do this by engaging their students in practical activities.

Familiarity with data collection and recording procedures. In Nigeria, science practical work involves making observations, collecting data, and recording and interpreting data for the purpose of making conclusions. There is usually a sequence of definite procedures to follow where observations precede other stages/processes. However, to appreciate that there is no standard scientific method by which all science is done, students need to understand that scientific investigation can follow more than one method, and experimental results or conclusions may still be consistent under certain conditions. This is one of the tenets of the nature of science, among others, and a goal of science learning that should inform practical work in school science.

Ability to draw inferences from observations. In science, it is not sufficient to merely make observations. The implications of observations, such as changes in colour, viscosity, states of matter, sound, odour and so on, are presented as conclusions or inferences. Students are taught the implications of observations in specific experiments and these observations and their corresponding implications are documented in science practical textbooks and guides which are used during practical exercises. The fact that students can refer to practical textbooks outside the laboratory affords them the opportunity to familiarize themselves with the indications of specific observations in laboratory practical work. This is further consolidated with their practical experiences in the laboratories. Hence, their ability to draw inferences from observations is strengthened.

Promoting teamwork among students. In several science practical activities, students work in groups of two or three and the success of such activities depends on the ability of the students to work together as a team. When students work together as a team, they do not only learn cooperatively; they engage in social interaction. A more recent purpose of school science in many countries that is beginning to be reflected in curriculum statements is the concept of employability of graduates and high school leavers who are equipped to solve novel problems, communicate effectively, work as a team, and apply many more interpersonal skills while using their scientific knowledge and expertise (e.g. Millar, 2010; Sarkar, Overton, Thompson, & Rayner, 2016). Although, practical work create opportunities for students to work in groups, peer review/ discussion of experimental results among students rarely take place. Instead, the teacher takes centre stage to 'explain away' students' observations and possibly credit the discrepancies in students' results to experimental error. In such situations, we do not see how the current practice of practical work in school science promotes effective communication and engage students in argument drawn from evidence.

These roles reflect the current practice of practical work in Nigerian school science. Apart from the challenges that plague practical work in school science, which will be considered in the next section, we do not yet see in practice the role of practical work as promoting students' understanding of the nature of science. Practical work in science is overly "cookbook" in nature, requiring students to follow instructions without thinking about what they are doing. It is questionable whether such practical work can really lead to meaningful learning.

Challenges of science practical work in Nigerian schools

A myriad of challenges confronts science practical work in Nigerian schools. In an attempt to unravel the factors responsible for the dismal performance of Nigerian students in School Certificate Science subjects, many studies have revealed factors associated with the implementation of practical science as major contributors to students' poor performance (Achimugu, 2012; Afemikhe, Imobekhai & Ogbuanya, 2018; Agogo, 2009; Alamu, 2012; Ogom, 2000). Some of these factors are discussed as follows:

Complete lack of laboratories. In many cases, laboratories are found to be unavailable in some schools. Hence, teachers have to make use of the normal school classrooms as makeshift laboratories. This makes the work of the teacher very difficult. The absence of a laboratory building is a strong indication that laboratory equipment and reagents may not also be available or if available will not be properly secured. Where some equipment and reagents are available, they are often stored in the School Principal's office for safety. The lack of laboratory often discourages science teachers from doing practical work, as it requires extra effort. A physics practical requiring the use of power may not be feasible in a makeshift laboratory. Laboratories are constructed on the basis of certain specifications, which are even peculiar to respective science subjects. For example, a biology practical activity may not be successfully conducted in a physics laboratory.

Poor power supply. In the modern-day world where scientific measurements are made using digital technologies that are powered by electricity, an unstable power supply is a hindrance to the proper conduct of science practical, especially those instruments for which batteries could not be used as an alternative source of power. In an experimental set-up where measurements have to be conducted using the old-fashioned analogue instruments, the results generated may not be accurate and precise. In Nigeria, communities are not connected to the national power grid. In such cases, certain practical work that requires electricity may be impossible except for schools that can afford to use power plants.

Inadequate laboratory equipment, reagents and consumables. Laboratory facilities/materials and consumables that have to be purchased from time to time are generally expensive. In many instances, some of the facilities are not readily available and where they are available, they are often grossly inadequate to cater for the large number of students in the science laboratory. In this situation, teachers are compelled to resort to demonstration of practical activities, leaving students to only observe the teachers' demonstration of scientific concepts/ theories. As a result, the definition of practical work as "any teaching and learning activity in which the students... manipulate the objects or materials..." becomes unachievable (Millar, 2010, p. 109).

Large class size. Many private and public schools in Nigeria have become over-populated. The national policy on education stipulates that the teacher-pupil/student ratio in a class should not exceed 1:40. This is no longer possible as classrooms, and the laboratories by extension, cannot accommodate the teeming population of students for science practical work. Teachers have to resort to postponing the practical lessons in order to make plans such as dividing students into groups where they take turns to do the practical activities. This often leads to the disruption of practical activities. Another resultant effect of mass education is the sharing of equipment and reagents by students. In such situations, only a few students participate in the practical tasks while others simply watch. This does not provide opportunities for students to acquire manipulative skills that are expected to be a critical part of the students' learning outcomes.

Poorly qualified science teachers. Many science teachers are not professionally trained and hence lack the requisite knowledge and skills for conducting science practical with students. Laboratory practice is part and parcel of science teacher training programmes in Nigeria and a non-professional may not be able to do this properly unless they makes some extra efforts to develop themselves. Professionalization of the teaching profession is an on-going process in Nigeria, and at the moment there is a shortage of professionally trained science teachers.

Inadequate and unqualified laboratory personnel. Laboratory technologists are the personnel trained and qualified to work in the science laboratories. They are also meant to be assisted by qualified personnel referred to as laboratory assistants. In most schools, there are neither qualified laboratory technologists nor laboratory assistants. Hence, science teachers double as technologists. In Nigeria, a common practice in most secondary schools is the handling of laboratory activities by poorly qualified personnel under the supervision of the science teacher. The role of the laboratory technologists has been undermined and this is one of the reasons why the laboratories are poorly managed. When the services of laboratory technologists who are trained to repair and maintain equipment are not readily available, the durability of laboratory facilities/ equipment may not be guaranteed, as there is no assurance that the facilities will be properly maintained.

The teaching of science practical separately from theory. For proper assimilation and concretization of scientific knowledge, the teaching and learning of science theory and practical should go together concurrently. In other words, when a teacher teaches electrolysis, the theoretical aspects of the topics must be complemented with practical demonstrations and practice simultaneously. This facilitates better assimilation and understanding of facts, principles and theories.

This is why it is ideal that science lessons should normally be held in the laboratories. The idea of teaching theoretical aspects of science in the classroom and then devoting a few periods for practical work after some time does not augur well for proper understanding, and leads to a disconnect between the theory and practical.

These challenges are not new, and neither are they peculiar to Nigerian schools alone. In actual fact, some of them have been around for decades. About two decades ago, Alebiosu (2000) and Onipede (2003) reported on the state of facilities in Nigerian school laboratories and noted that the facilities were inadequate. Jones (1990) had earlier surveyed the state of school laboratories in some selected African countries and found that as much as 45% of the schools had insufficient laboratories. It is particularly worrisome that nearly 30 years later, these challenges have persisted and there seems to be no available empirical evidence to show any significant progress. This calls for serious reflection by all stakeholders in science education in Nigeria.

Concluding comments

The aim of this chapter was to discuss the role of practical work in science teaching and learning, as it is currently undertaken in secondary school science in Nigeria. What we see emerging from the narrative of the nature of practical work used in school science teaching is the widespread use of "cookbook" experiments or recipe-style tasks that serve the academic function of what Osborne (2014) described as "illustrating or verifying the phenomenological account of nature" (p. 178). Although the nature of practical work or laboratory activities in secondary schools allows students to manipulate objects and materials, what seems to play out in school science is that such activities are usually designed to reflect what the teachers anticipate or expect the students to produce in examining the phenomenon or phenomena under investigation. This raises the question of whether school science practical work is effective in helping students to establish links between what they do and the scientific ideas behind the scientific investigations. In cases where this connection is missing, students may not be able to use these scientific ideas to make sense of their observations or evidence.

Practical work in itself has the potential to provide opportunities to link science concepts and theories with observations of phenomena (Hofstein & Lunetta, 2004). While the students can be engaged with the doing of practical work, it is important that teachers provide some sort of scaffolding to enable them to see what the teacher expects as learning outcomes and to help them reflect on the observations they have made. The Predict-Observe-Explain (POE) approach developed by White and Gunstone (1992) may help students develop conceptual understanding of the material world. Abrahams and Michael (2012) describe this approach as an effective way to get students to think about their practices before and after practical tasks.

References

Abimbola, I. O. (2009). Scientific literacy and the African worldview: Implications for sustainable development of science education in Africa. In K. Opoko-Agyemang (Ed.), *Culture, science and sustainable development in Africa* (pp. 344–362). Cape Coast: University Press.

Abrahams, I. (2009). Does practical work really motivate? A study of the affective value of practical work in secondary school science. *International Journal of Science Education*, 31(17), 2335–2353.

Abrahams, I. (2011). *Practical work in secondary science: A minds-on approach*. London: Continuum.

Abrahams, I., & Michael, R. (2012). Practical work: Its effectiveness in primary and secondary schools in England. *Journal of Research in Science Teaching*, 49(8), 1035–1055.

Abrahams, I., & Millar, R. (2008). Does practical work really work? A study of the effectiveness of practical work as a teaching and learning method in school science. *International Journal of Science Education*, 30(14), 1945–1969.

Achimugu, L. (2012). Strategies for effective conduct of practical chemistry works in senior secondary schools in Nigeria. *Journal of the Science Teachers Association of Nigeria*, 47(1), 126–136.

Afemikhe, O. A., Imobekhai, S. Y., & Ogbuanya, T. C. (2018). *Assessment of valid science practical skills for Nigerian secondary schools: Teachers' practices and militating factors*. Proceedings of the 41st annual conference of the International Association for Educational Assessment.

Agogo, P. O. (2009). *Topics in primary science education in Nigeria*. Makurdi: Azaben Press. Cambridge Assessment.

Alamu, A. A. (2012). The state of science and technology infrastructure in secondary schools in Nigeria. *Journal of Educational and Social Research,* 2(8), 56–66.

Alebiosu, K. A. (2000). Effects of two instructional methods on senior secondary school students' perceptions of the difficulty in learning some chemical concepts and their achievement gains. *Journal of Education Foundations Management,* 1, 55–64.

Anderson, D. (2012). *Teacher knowledges, classroom realities: Implementing sociocultural science in New Zealand Year 7 and 8 classrooms*. Unpublished doctoral dissertation, Victoria University of Wellington, New Zealand.

Department forEducation (2011). *Written evidence submitted by the Department for Education to the House of Lords Science and Technology Committee*. Retrieved from www.parliam ent.uk/business/committees/committees-a-z/commons-select/science-and-technolo gy-committee/inquiries/school-science/

Donnelly, J., Buchan, A. S., Jenkins, E. W., & Welford, A. G. (1993). *Policy, practice and professional judgement: School-based assessment of practical science*. Driffield: Studies in Education.

Hodson, A. (1992). Assessment of practical work: Some considerations in philosophy of science. *Science & Education*, 1, 115–144.

Hodson, D. (1991). Practical working science: Time for a reappraisal. *Studies in Science Education*, 19, 175–184.

Hofstein, A., & Lunetta, V. N. (2004). The laboratory in science education: Foundations for the twenty-first century. *Science Education*, 88(1), 28–54.

Hofstein, A., Kipnis, M., & Abrahams, I. (2013). How to learn in and from the chemistry laboratory. In I. Eilks & A. Hofstein (Eds.), *Teaching chemistry: A studybook* (pp. 153–182). Rotterdam: Sense Publishers.

Hume, A., & Coll, R. (2008). Student experiences of carrying out a practical science investigation under direction. *International Journal of Science Education*, 30(9), 1201–1228.

Iyiola, A. O., & Wilkinson, W. J. (1984) A study of the biological science curriculum for secondary schools in Nigeria. *Research in Science and Technological Education*, 2(2), 139–152. doi:10.1080/0263514840020206

Jegede, O. J. (1988). The development of the science, technology and society curricula in Nigeria. *International Journal of Science Education*, 10(4), 399–408.

Jones, E. (1990). Teacher provision in the sciences. *Journal of Science Education, 140*, 27–37.

Jones, T. J. (1922). *Education in Africa: A study of west, south and equatorial Africa*, African Education Commission, Phelps-Stokes Fund. New York: Wiley.

Millar, R. (1999, August). *Understanding how to deal with experimental uncertainty: A 'missing link' in our model of scientific reasoning?* Paper presented at the conference of the European Science Education Research Association (ESERA), Kiel, Germany.

Millar, R. (2004). The role of practical work in the teaching and learning of science. Paper prepared for the *Committee: High school science laboratories: Role and vision*. Washington, DC: National Academy of Sciences.

Millar, R. (2010). Practical work. In J. Osborne & J. Dillon (Eds.), *Good practice in science teaching: What research has to say* (2nd ed.) (pp. 108–134). Maidenhead: Open University Press.

Millar, R. (2011). Reviewing the national curriculum for science: Opportunities and challenges. *Curriculum Journal*, 22(2), 167–185.

Moeed, A., & Anderson, D. (2018). *Learning through school science investigation: Teachers putting research into practice*. Singapore: Springer.

Moeed, H. A. (2010). *Science investigation in New Zealand secondary schools: Exploring the links between learning, motivation and internal assessment in Year 11*. Unpublished doctoral thesis, Victoria University of Wellington, New Zealand.

Needham, R. (2014). The contribution of practical work to the science curriculum. *School Science Review*, 95(352), 63–69.

Nigerian Educational Research and Development Council (NERDC) (2009). *Senior secondary education curriculum: Chemistry for SS1–3*. Yaba, Lagos: NERDC Press.

Ogom, O. I. (2000). *New approach to chemistry practical for senior secondary school*. Otukpo: Dorcas Publishers.

Ojerinde, D. (2011). *Public examinations in Nigeria*. New Delhi: Melrose Books & Publishing Limited.

Onipede, H. (2003, January 2). National development hinges on quality education. *The Comet*.

Osborne, J. (1998). Science education without a laboratory? In J. Wellington (Ed.), *Practical work in school science: Which way now?* (pp. 156–173). London: Routledge.

Osborne, J. (2014). Teaching scientific practices: Meeting the challenge of change. *Journal of Science Teacher Education, 25*(2), 177–196.

Ramnarain, U., & Kibirige, I. (2010). Learning through investigations: The role and place of scientific investigations in school science. In U. Ramnarain (Ed.), *Teaching scientific investigations* (pp. 1–23). Cape Town: MacMillan.

Sarkar, M., Overton, T., Thompson, C., & Rayner, G. (2016). Graduate employability: Views of recent science graduates and employers. *International Journal of Innovation in Science and Mathematics Education, 24*(3), 31–48.

Toplis, R. (2012). Students' views about secondary school science lessons: The role of practical work. *Research in Science Education,* 42, 531–549. doi:10.1007/s11165-011-9209-6

Wellington, J. (1998) (Ed). *Practical work in school science: Which way now?*London: Routledge.

West African Examinations Council (WAEC) (2005). *Regulations and syllabuses for the West African school certificate examination (WASSCE).* Accra, Ghana: WAEC

White, R. T. (1988) *Learning science.* Oxford: Basil Blackwell.

White, R., & Gunstone, R. (1992). *Probing understanding.* London: Falmer Press.

Chapter 5

The pedagogical orientations of Malawian science teachers towards practical work

Dorothy C. Nampota, Nellie M. Mbano and Bob Maseko

Introduction

Practical work has long been placed at the centre of school science teaching globally, to the extent that Solomon (1980) wrote:

> Science teaching must take place in the laboratory; about that at least there is no controversy. Science simply belongs there as naturally as cooking belongs to the kitchen and gardening in the garden.
>
> (Solomon, 1980, p. 13)

Despite the widespread use of practical work as a teaching and learning strategy in school science, some science educators have raised questions about its effectiveness. Hodson (1991), for example, claims that: 'As practised in many schools it [practical work] is ill-conceived, confused and unproductive. For many children, what goes on in the laboratory contributes little to their learning of science' (p. 176). This points to the need to have a clear understanding of what practical work is and how it should be taught. Interestingly, despite its widespread use, there is no consensus on what is meant by practical work, resulting in variations in practice in different laboratories and countries. The lack of consensus has further been exacerbated by the introduction of new related concepts of 'investigations' and 'inquiry'. One of the most used definitions of practical work, however, is that proposed by Lunetta, Hofstein, and Clough (2007), who defined practical work as:

> ...learning experiences in which students interact with materials or with secondary sources of data to observe and understand the natural world (for example aerial photographs to examine lunar and earth geographic features; spectra to examine the nature of stars and atmospheres; sonar images to examine living systems). (p. 394)

The emphasis is on interaction with materials of some sort, including secondary data, in order to understand phenomena. Other similar definitions include that earlier made by Millar (2004, p. 1), who defined practical work as 'any teaching and learning activity which at some point involves the students in observing or manipulating the objects and materials they are studying', although this excludes the aspect of secondary data. A group of researchers working as a Science Community Representing Education (SCORE) in the United Kingdom, which is a partnership of six UK organisations including the Association for Science Education, provided further clarification on activities that comprise practical work. In a survey of 1103 teachers on their understanding of practical work, the researchers found that practical work activities can be put into three broad categories: core activities, directly related activities and complementary activities (SCORE, 2008). Core activities include investigations, laboratory procedures and techniques and fieldwork; directly related activities include designing and planning investigations, data analysis using ICT, analysing results, teacher demonstrations and experiencing phenomena; complementary activities include science-related visits, surveys, presentations and role play, simulations including use of ICT, models and modelling, group discussion and text-based group activities. The SCORE report, however, indicated that of the three categories, practical work in science primarily consists of the core activities and the directly related activities. The complementary activities were seen as necessary in supporting the development of conceptual understanding in science through practical work.

The inclusion of 'investigations' in the core activities broadens the scope of practical work beyond the definition by Lunetta et al. (2007) and Abrahams and Millar (2008), where the emphasis was on the student interacting with materials or analysis of secondary data in order to explore phenomena. Investigations are more comprehensive and include several processes: a) developing an idea for investigation, b) designing and planning investigations, c) carrying out the investigation and d) evaluating and interpreting the evidence. The definitions by Lunetta et al. and Abraham and Millar are more concerned with processes (c) and (d). Combining the core activities and directly related activities, the activities to be categorised as practical work would include those summarised in Table 5.1.

Table 5.1 Activities that comprise practical work

Stage	Activities
1. Developing an idea for investigation	Identifying questions to investigate
2. Planning the investigation	Designing procedures
3. Carrying out the test	Handling materials Taking measurements Recording data
4. Analysing data	Transforming data in different forms Drawing relationships between variables
5. Evaluating and interpreting evidence	Drawing conclusions

Roles of practical work

Despite the lack of consensus in defining practical work, the literature (see Abrahams, 2011; Wellington, 1998) is clear that practical work plays a critical role in the teaching and learning of science. While learning from pre-service teachers on how they understand the role or purpose of practical work, Wellington (1998) described and compressed teachers' ideas about the role of practical work into three main arguments: the *cognitive argument, the affective argument* and the *skills development argument*. Under the cognitive argument, even though other science educators have argued that practical work is not the best way of teaching 'theory', since theory is about ideas and not 'things', other researchers (see Abrahams & Reiss, 2012) still contend that practical work significantly improves students' understanding of science. It is believed that through visualisations of the laws and theories accorded through practical work, there is a promotion of conceptual understanding. Under the affective argument, it is argued that practical work not only motivates the students; it also excites the students. This excitement comes about from the fact that students are given a platform where they observe science in action. On the other hand, the skills argument advances the understanding that practical work not only excites the students; it also instils critical science skills in the learners. These skills range from manipulative skills or dexterity skills to higher-order transferrable skills such as observation, measurement, prediction, as well as inferential skills (Wellington, 1998).

Furthermore, Abrahams (2011) described five roles of practical work, most of which fit within the three arguments proposed by Wellington. The first is that practical work enhances the learning of scientific knowledge, which is linked to Wellington's cognitive argument. In a study where the role of practical work was investigated, Hewson and Hewson (1983) found that practical work improves students' understanding of concepts, theories as well as laws in science. The authors used two groups of students to achieve their aim: a control group which was exclusively taught using traditional methods and a non-controlled group that used student-centred methods. Secondly, the aim of practical work, according to Abrahams (2011), is that it enhances student motivation, which links to the affective argument. In a survey of over 1,400 students on the teaching methods that they find enjoyable, Cerini, Murray, and Reiss (2003) found that 71% of the students chose 'doing an experiment in class'. However, there is also evidence that other studies have failed to replicate this finding. For instance, in a study that sought to investigate the affective value of practical work, Abrahams (2009) reported that while practical work generated short-term engagement in science lessons, it was found to be ineffective in motivating students. Thirdly, Abrahams (2011) argues that practical work helps to teach laboratory skills. Even though the term 'skills' in this case is always challenging, it has been understood that some of these laboratory skills include handling apparatus, taking measurements, problem-solving, general intellectual development, observing and classifying, just to mention a few. This links to the skills development argument of Wellington (1998). Fourthly, Abrahams claims that practical work helps students to develop 'scientific attitudes' such as open-mindedness, objectivity and willingness to suspend judgement. What seems to be coming out of this is that practical work implicitly helps students to learn about the nature of science (NoS) and how science ought to be conducted. Science demands that students or practitioners are objective and have an open mind – this means they should be able to expect the unexpected. Furthermore, science demands that those practising science should be aware of the nature of scientific knowledge as well as how that particular knowledge is generated. All these are aspects of the NoS; hence it might be argued that practical work contributes to the learning about science. Woolnough and Allsop (1985) posit yet a fifth role of practical work, which is to enable students to get a 'feel for phenomena'. A 'feel for phenomena' means to have first-hand experience of the phenomena such as observing the behaviour of a current in an electric circuit or observing the colour change in a titration experiment.

Practical work in Malawi

The education system in Malawi follows an 8–4–4 system with eight years of primary, four years of secondary and four years of university education. Practical work is taught throughout the four years of secondary schooling, which is years 9–12 of education. The four years of secondary education are known as 'Forms' and classified as Junior Secondary (Forms 1 and 2) and Senior Secondary (Forms 3 and 4). In keeping with the global picture, practical work plays a central role in the teaching and learning of Biology, Chemistry and Physics.

The importance of practical work has been re-emphasised in the new secondary school curricula rolled out in 2016 (Government of Malawi, 2013a, p. ix), which reflects a shift from the use of integrated and general science syllabi, which required less practical work, to separate sciences that require a lot of practical work. The rationale for the three separate science subjects of Biology, Chemistry and Physics all reflect the importance of practical work in the development of skills necessary to solve everyday problems (own emphases):

> The student will be able to apply thinking, study, problem-solving and *investigative skills and techniques* to solve problems in everyday life.
>
> (Government of Malawi, 2013a, p. xi)

> Through the *investigative approach*, chemistry equips students with essential skills for effective communication of scientific information, problem-solving and pursuit of further education.
>
> (Government of Malawi, 2013b, p. xi)

Physics helps students to become more scientifically literate, i.e. it enables them to *think critically and creatively based on explanations developed and evaluated from experiments and models*. The subject will, therefore, help students to develop a *scientific mind/view* necessary for identifying and solving current and emerging/new scientific issues.

(Government of Malawi, 2013c, p. xi)

An analysis of the Chemistry syllabus shows that there is a significant difference in the structure of practical work at the junior and senior level of secondary school education. While the focus of practical work in junior syllabi (Forms 1 and 2) is on conducting investigations (taking measurements, safety), collecting (graduating instruments, accuracy, safety), and analysing data and drawing conclusions (presenting data in different forms, interpreting data), the focus in senior syllabi (Forms 3 and 4) is on designing investigations (where students are expected to develop a procedure) in addition to conducting investigations, collecting and analysing data and drawing conclusions (Government of Malawi, 2013a). Similarly, both the junior and senior Physics syllabi are explicit about all the stages of practical work – Asking questions (identifying a problem, hypothesising), Designing investigations (identifying variables, deciding the type of data to collect); Conducting investigations (using basic instruments, taking readings); Collecting and analysing data (taking readings, organising data, plotting graphs); Drawing conclusions (drawing conclusions, relating conclusions to hypothesis) (Government of Malawi, 2013c).

The key phrases that describe the nature of practical work in these definitions are scientific methods, techniques and materials, investigative skills and techniques, test theories, analyses and evaluate scientific data from observations and experiments. These definitions fall within the description of practical work provided by (SCORE, 2008). However, whether or not the whole set of activities and processes written in the intended curriculum are practised in a practical lesson in the laboratory depends on the teachers.

Research has shown that teacher practices are influenced by a number of factors including beliefs, teacher preparation through pre-service training and contextual factors (Ramnarain, Nampota & Schuster, 2016) amongst others. In a study of teacher pedagogical orientations in science teaching conducted in Malawi and South Africa, Ramnarain et al. (2016) found that teachers from disadvantaged schools tended to lean more towards direct forms of instruction while teachers at relatively privileged schools are more inclined to adopt inquiry forms of instruction. The rather extreme Direct Didactic orientation is prevalent amongst teachers from disadvantaged community day secondary schools, which have limited availability of laboratory equipment and consumables and are often characterised by large classes of over 60 students.

Another factor that influences instruction, at least in most developing countries, is examinations. Teachers have argued that they teach towards examinations because of the competitive nature of the education system that has a pyramidal shape with a broad primary school setting and narrowing in secondary and further narrowing at tertiary level. In Malawi however, deliberate initiatives have been put in place in order to ensure that practical work is taught in schools. With regard to examinations, there is a separate practical examination paper for each of the science subjects at the school leaving certificate level, which is Form 4. The practical examination involves students conducting investigations whose questions or aims are provided; recording data, usually in a given format; and providing answers to questions that are provided in the examination paper to help students draw conclusions. While students undertake these practical work tasks individually over a specified time, the process of conducting the investigation is not assessed. Rather, once the students have finished providing answers to the given set of questions, the papers are marked and therefore there is a focus on assessing the product. Moreover, this type of assessment is conducted once at the end of Form 4 and not in the other years of secondary schooling.

Having discussed the influence of contextual factors and the assessment practices for practical work, questions arise as to whether or not the intentions about practical work as captured by the written Biology, Chemistry and Physics syllabi are translated into practice in the laboratory in Malawi. This necessitated the present study, which was explicitly guided by the following research questions:

a   What are the pedagogical orientations of Malawian science teachers towards practical work?
b   How does teacher understanding of the role of practical work influence the pedagogical orientations?

A framework for analysing teacher pedagogical orientations

A critical dimension in analysing the nature of practical work in the school laboratory is the pedagogical orientation assumed by teachers. Pedagogical orientation has been theorised as a component of pedagogical content knowledge (PCK) by Magnusson, Krajcik, and Borko (1999) who used the term 'orientation' to refer to 'teachers' knowledge and beliefs about the purposes and goals for teaching science at a particular grade level' (p. 97). Law (2009) maintains that the three aspects most indicative of the pedagogical orientation of the teacher are the curriculum goals, the roles played by the teacher as reflected by their teaching practices, and the roles played by students in their learning practices.

A research team at Western Michigan University introduced the concept of a *science teaching orientation spectrum* (Cobern et al., 2010) in relation to concept formation. In this spectrum, teaching approaches are classified into four main types: two variants of direct instruction and two variants of inquiry instruction. The two variants of direct instruction are called Direct Didactic (DD) and Direct Interactive (DI) approaches. A teacher who assumes a DD approach presents the science concept or principle directly to the students, explains, and illustrates with examples and/or demonstrations. Students apply this knowledge to questions and problems. A DI orientation similarly entails direct exposition, but this is followed by a student activity based on the presented science, e.g. hands-on practical verification of a law. In both of these approaches, the teacher has more control of the teaching situation and student autonomy is low. Philosophically, this depicts a positivist view of science as ready-made knowledge. The two variants of inquiry instruction, on the other hand, present the philosophical view of science-in-the-making through Guided Inquiry (GI) and Open Inquiry (OI). In adopting a GI orientation, the teacher plans an activity where students explore a phenomenon or idea, and the teacher guides them to develop the desired science concept or principle. In OI, students explore a phenomenon or idea on their own, minimally guided, and devise ways of doing so, after which students present what they have done and discovered. The teacher facilitates but does not intervene more than necessary, and the emphasis is on the inquiry process. In the inquiry approaches, teacher control decreases and students' autonomy increases.

Adapting the science teaching orientation spectrum to practical work, a teacher who assumes a DD approach presents or demonstrates practical activities directly to the students, explains the procedure, provides a table for recording data and explains the conclusion. A DI orientation similarly entails direct exposition to an experiment, e.g. hands-on practical, but this focusses on verification of a law that the student has already learnt. In adopting a GI approach, the teacher plans a practical activity where students explore a phenomenon or idea, whose answers they did not have prior knowledge on, and the teacher guides them to develop the desired science concept or principle or explanation. The students devise their own ways for recording data and coming up with a relationship between the variables. In OI, students explore a phenomenon or idea on their own, devise questions and ways of exploring the phenomena, after which students present what they have done and discovered.

Methodology

Study design

The study followed a descriptive survey design with the aim of describing teacher pedagogical orientations as they implement practical work in the laboratory. Rather than soliciting the perceptions of teachers on practical work through questionnaires, we explored how practical work is actually conducted in classrooms through observations. The observations were augmented by teachers' explanations of their actions through post-lesson interviews and students' perceptions of teacher pedagogical orientations through questionnaires.

The sample

The sample consisted of 18 teachers purposefully selected from six conventional[1] secondary schools in Zomba district. The choice of conventional schools was important in order to exclude some contextual factors that have been found to influence teacher practices (Ramanrain et al., 2016)[2]. Permission was sought from the Education Division offices in charge of secondary education, the school head teachers and the teachers themselves. The teachers were asked for their willingness to teach a practical lesson at an appropriate time. The characteristics of the teachers and the observations made are summarised in Table 5.2.

A total of 655 students whom the target teachers were teaching on the day of the classroom observation also took part in the study. Three out of the 655 students did not indicate their gender; out of the 652 students who indicated their gender, 303 (46.5%) were male while 349 (53.5%) were female. The relatively larger number of females is explained by the fact that one of the six target schools was a girls-only school. The students were distributed as follows for Form 1, Form 2, Form 3 and Form 4, respectively: 90 (13.7%); 31 (4.6%); 254 (38.8%); and 279 (42.6%). This shows that the majority of the students were from the senior classes and therefore had a lot of experience of practical work.

Methods

Data were collected through three methods that allowed triangulation of data collected: classroom observations, post-lesson interviews and student questionnaire. Classroom observations were used to capture first-hand information on the pedagogical orientation of the teachers while the post-lesson interviews sought to seek clarifications from the teacher on the pedagogical orientation used. The student questionnaire was used to explore the students' perception of the teachers' pedagogical orientation.

Table 5.2 Characteristics of teachers and the lessons observed

Characteristic	Number
No. of lessons	18
No. of teachers	18 (15 Male, 3 Female)
Qualifications	17 B.Ed. and 1 Dip Ed
Classes observed	Form 4: 12 lessons, Form 3: 3 lessons, Form 2: 1 lesson and Form 1: 2 lessons
Subjects observed	Biology: 3 lessons, Chemistry: 9 lessons and Physics: 6 lessons
Teaching experience	2–22 years

a) Classroom observations

Classroom observation was the primary data collection technique for the study. The distinct feature of observation as a research process is that 'it offers an investigator the opportunity to gather "live" data from naturally occurring social situations' (Cohen, Manion, & Morrison, 2007, p. 369). This has an advantage over reported data, which tends to give perceptions. Observation studies have several dimensions, such as whether they are structured or unstructured, and whether they are participant or non-participant. In this study, non-participant structured observation was used. A structured observation schedule was used. This is an adaptation of an instrument used to measure the principles of scientific inquiry developed and trialled by Campbell, Abd-Hamid, and Chapman (2010), and is closely linked to the framework of the pedagogical orientation. The instrument has five categories which closely mirror the definition of practical work presented earlier: a) asking questions/framing research questions, b) designing investigations, c) conducting investigations, d) collecting data and e) drawing conclusions. Each of the five categories was distinct and while some had indicators depicting DD, DI, GI and OI orientations, others had a subset of the orientations as shown in Table 5.3. In total, the observation schedule had 18 indicators either focussed on the activities of the teacher or of the students, depicting the DD and OI continuum. A timeframe was added to the observation schedule so that the observer scored the categories every two minutes using a timer. Several categories could be scored in each two-minute segment as they occurred (see Table A.1 in the chapter appendix).

In order to test the reliability of the instrument, three observations were jointly made by the researchers and discussions afterwards focussed on interpretation of the categories and the mechanism of recording the observations in order to test its reliability. Reliability was measured using the simple formula of (Number of times two observers agree/Number of possible opportunities to agree) × 100%, which resulted in a score of 99.1%. The weakness of the schedule, which might have contributed to the high score, was that 4 of the 18 indicators were missing in all the lessons observed.

b) Post-lesson teacher interviews

All the 18 teachers were interviewed after their laboratory lesson was observed in order to obtain an explanation for the particular orientation exhibited. Examples of questions included: what did you intend your students to learn in this practical? Why did you provide the students with a table for recording results (depending on what the teacher had done)? The post-lesson interviews also sought answers to the second research question on the teacher's understanding of the roles of practical work and how it should be implemented in the laboratory. The post-lesson interviews were necessary because 'when an interviewee is aware that the interviewer has observed the practice being

Table 5.3 Indicators and pedagogical orientation

Activity/Indicator	Orientation
A. Asking questions/framing research questions	
A1. Teacher gives students questions to investigate	DD
A2. Teacher asks students to refine given questions to determine direction of the practical	DI
A3. Teacher asks students to formulate questions to investigate	OI
A4. Teacher discusses student research questions to determine the direction of the practical	GI
B. Designing investigations	
B1. Teacher gives step-by-step instructions before students conduct investigations	DD
B2. Students engage in a critical assessment of the procedures given by the teacher	DI
B3. Students design their own procedures for investigations	OI
C. Conducting investigations	
C1. Students conduct procedures of an investigation	DI
C2. Teacher conducts the investigation without involvement of the students	DD
C3. Teacher conducts investigations with students' involvement	DI
D. Collecting data, analysis and presentation	
D1. Teacher provides a way for recording data	DD
D2. Students design a way of recording data	OI
D3. Students determining how to present the data	OI
D4. Teacher tells students how to present the data	DD
E. Drawing conclusions	
E1. Teacher draws conclusions from the practical	DD
E2. Students state their own conclusions for investigations	OI
E3. Students connect conclusions to scientific knowledge	OI
E4. Students justify their conclusions	OI

discussed, responses are more effectively anchored to realities, and less likely to be 'rhetorical' in nature' (Abrahams & Millar, 2008, p. 1950). The interview guide is provided in the chapter appendix.

c) Students' questionnaire

The lesson observation schedule with the 18 indicators was administered to students who took part in the lessons observed during the study, and therefore is referred to as a student questionnaire. The questionnaire had the same indicators and the student responses were presented in a Likert-scale format showing the options *never, sometimes* and *almost always* (see the chapter appendix). The purpose of the questionnaire was to complement the insights gained from the teacher observation and interviews. The questionnaire was administered in a classroom setting with the researcher reading each statement out loud, explaining it clearly and giving the students time to answer by ticking the appropriate box of *never, sometimes* and *almost always*. The total time taken to complete the questionnaire was about 10 minutes. In total, 655 students answered the questionnaire; three students did not declare their gender. Out of the 652 students that had their gender entered, 303 were male, and 349 were female. Across the grades, the number of students was 90, 31, 254 and 249 for Forms 1, 2, 3 and 4 respectively. The focus on the senior grades of Form 3 and 4 was deliberate as the targets were experienced teachers.

Data analysis

The classroom observation data and student questionnaire data, which sought to answer the question of teacher pedagogical practices, were analysed using SPSS software to generate descriptive statistics. The indicators in the classroom observation schedule and student questionnaire had already been coded as DD, DI, GI and OI. Data from the post-lesson interviews, particularly on the teachers' perceived role of practical work, were inductively analysed to identify emerging themes. A qualitative data management software program – NVivo 12 – was used to manage the qualitative data analysis process. To facilitate inductive analysis, we converted the transcripts into Word format, which were then cleaned to ensure uniformity, and later imported into the NVivo 12 software. An iterative read and re-read procedure was used to identify emerging themes under which the interview data were coded.

Results

a) Teacher pedagogical orientations

Teacher pedagogical orientations towards practical work were assessed through classroom observations and the students' questionnaire. The findings from the classroom observations are summarised in Table 5.4.

Table 5.4 shows that under category A of practical work which focussed on framing questions for investigation, the only practice observed was that the teacher gave students questions to investigate. This was evident in 12 out of the 18 lessons observed, meaning that in some lessons no questions were posed before students undertook the investigations. Under category B, which focussed on designing investigations, in all the lessons observed the teacher gave step-by-step instructions (B1) and only two teachers allowed their students to do some critical assessment of the steps given. Under category C, which focussed on conducting investigations, all teachers involved their students in

Table 5.4 Teacher activities, average occurrence and number of lessons observed

Activity	Orientation	Average occurrence per lesson (n=18)	No of lessons
A1. Teacher gives students questions to investigate	DD	0.8	12
A2. Teacher asks students to refine given questions to determine direction of the practical	DI	0	0
A3. Teacher asks students to formulate questions to investigate	OI	0	0
A4. Teacher discusses student research questions to determine the direction of the practical	GI	0	0
B1. Teacher gives step-by-step instructions before students conduct investigations	DD	4.5	18
B2. Students engage in a critical assessment of the procedures given by the teacher	DI	0.1	2
B3. Students design their own procedures for investigations	OI	0	0
C1. Students conducting procedures of an investigation	DI	12.0	18
C2. Teacher conducting the investigation without involvement of the students	DD	0.4	1
C3. Teacher conducts investigation with students' involvement	DI	0.1	1
D1. Teacher provides a way for recording data	DD	1.8	14
D2. Students design a way of recording data	OI	0.4	5
D3. Students determine how to present the data	OI	0.4	2
D4. Teacher tells students how to present the data	DD	1.6	13
E1. Teacher draws conclusion from practical	DD	0.8	9
E2. Students state their own conclusions for investigations	OI	3.1	12
E3. Students justify their conclusions	OI	1.3	7
E4. Students connect conclusions to scientific knowledge	OI	0.8	4

conducting the investigation, although, at one point, one teacher conducted the investigation without the involvement of students. Under category D on collecting and analysing data, teachers in 14 out of the 18 lessons observed provided a way for recording data and even provided details on how students should represent the data. However, in a few (five) of the lessons, students designed their own way of recording data, and in two of such lessons, the students also determined how to present data on their own. Under category E focussing on drawing conclusions, students drew their own conclusions in 12 out of the 18 lessons observed although in some cases these were guided by the teacher. In six lessons, students justified their conclusions while in four lessons, students connected conclusions to scientific knowledge.

Figure 5.1 presents a graphical representation of the occurrence frequencies of the various activities summarised in Table 5.3. The figure shows that the activity that occurred most frequently in the practical lessons observed was the students conducting procedures of an investigation (C1). This activity occurred on average in 12 of the two-minute segments (24 minutes). The second most frequently occurring activity was B1, which is the teacher giving instructions for the practical activity, which on average occurred in nine minutes of a lesson. Students also spent a considerable amount of time drawing conclusions from the practical (E2) (an average of six minutes). Occurring less frequently were students giving a critical assessment of the procedure (B2), the teacher conducting an investigation without the involvement of the students (C3), and students designing their own way of recording results (D2) and representing data (D3). The categories related to students formulating questions to investigate or refining them were not observed in any of the lessons. Similarly, in no lesson were students asked to design their own procedures. It would seem that practical work focusses on hands-on experience

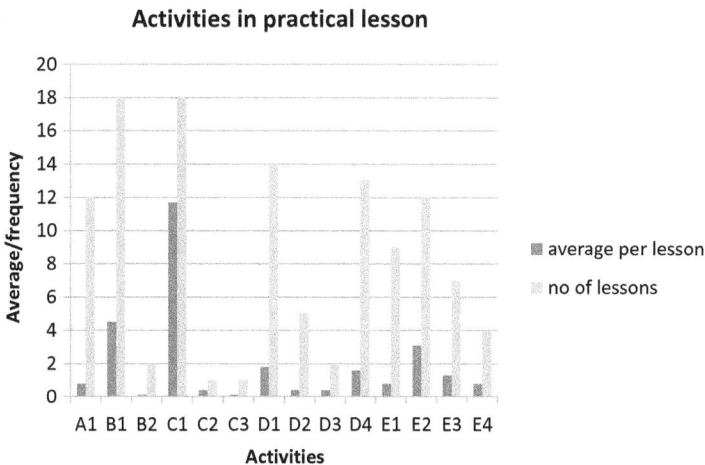

Figure 5.1 Average occurrence and frequency of practical work activities

and has very little minds-on experience. The teacher dominated the lessons by giving the question to be investigated and giving students step-by-step instructions, giving them instructions on how to record and present data. However, the students carried out the procedures. Although in a considerable number of lessons, students were given the opportunity to state their conclusion, there was little discussion to justify them or relate them to theory.

Using the framework of the pedagogical orientation, it is evident that for most of the practical work, DD pedagogical orientation dominates with instances of DI orientation, especially when students are conducting procedures of an investigation. The teachers visited groups while the students worked to check and ensure they were following the instructions, thereby augmenting the DD orientation. OI orientation was not observed at all, and while some aspects of GI orientation were evident in some lessons where students were drawing their own conclusions from an activity designed by the teacher, this was observed to a limited extent. In general, not much time was spent on discussing the questions, making plans of what to do or how to record data and how to present or analyse it.

The findings from the students' questionnaire, which had the same 18 statements, are similar to the classroom observation findings. For example under the formulation of questions for investigation (category A), about two-thirds (66.9%) of the students said that the teacher almost always gives them questions to investigate to which they are almost always (32.9%) and sometimes (45.5%) given an opportunity to discuss and refine the questions (see Figure 5.2). Nearly half of the students (47.9%) said that they are never given an opportunity to formulate questions for investigation and only 25.6% said their questions are never used as a basis for the practical activities they do. However, 30.7% of the students said they are

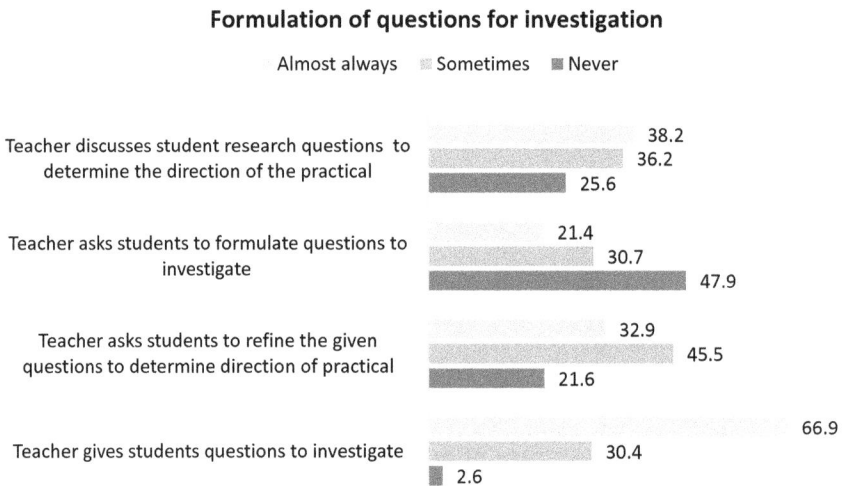

Formulation of questions for investigation

Almost always   Sometimes   ■ Never

Teacher discusses student research questions to determine the direction of the practical	38.2 / 36.2 / 25.6
Teacher asks students to formulate questions to investigate	21.4 / 30.7 / 47.9
Teacher asks students to refine the given questions to determine direction of practical	32.9 / 45.5 / 21.6
Teacher gives students questions to investigate	66.9 / 30.4 / 2.6

Figure 5.2 Formulation of questions

sometimes given opportunities to formulate questions which are almost always (38.2%) and sometimes (36.2%) used as a basis for the practical. If the students' questions are not used as a basis for the practical, then teacher control is still predominant and therefore leans more on the DD orientation. However, where the questions are refined and become a basis for the practical, then there is some thinking demanded of the students, therefore depicting a DI orientation.

It is interesting to note that teachers provide questions for investigation across all grades although, from an analysis of the curriculum, students should be taught how to formulate their own questions especially in Form 1 and Form 3 where the topic of scientific investigations is taught.

Under category B, which concerns designing investigations, Figure 5.3 shows that for the majority of the students (85.1%), investigations are designed by the teacher who almost always provides step-by-step instructions before the practical. 62.8% of the students said that they are never given an opportunity to design their own procedures for an investigation. This is true across all grades. Slightly over a third of students, however, said that they are almost always (37.1%) and sometimes (28.9%) given a chance to assess the procedures critically. Again, the trend was the same across grades.

Under conducting investigations (category C), only 12.4% of the students said that teachers conduct investigations on their own, meaning that students are involved in the process. 58.2% said that they always take part in conducting the investigation with the teacher, and 57.1% said they almost always conduct the procedures of the investigation. The findings, therefore, show that students are given hands-on experiences in the laboratory, following procedures that are designed by the teacher (see Figure 5.4).

Under category D on collecting and analysing data, the findings depict a DD orientation to practical work because the majority of the students (65.9%) said that the teacher almost always provides a way for recording data and 64.8% said

Designing investigations

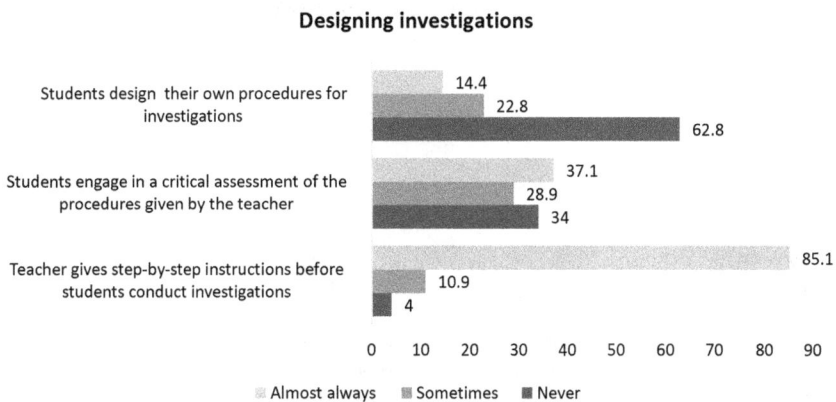

Figure 5.3 Designing investigations

Conducting investigations

■ Never   ▧ Sometimes   ▨ Almost always

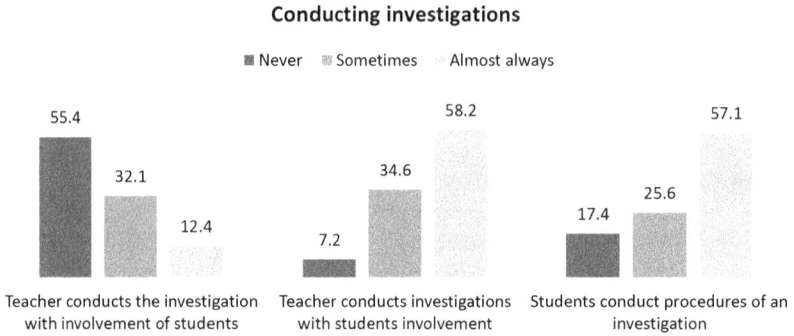

Figure 5.4 Conducting investigations

that they tell them how to represent the data in the form of graphs and other forms of analysis. 33.8% of the students said they are never given a chance to design their own way of recording data while 22.4% said they are never given the opportunity to decide how they are going to represent the data. This is true across the different grades, as shown in Figure 5.5.

Under category E on drawing conclusions, a fifth of the students said that the teacher never draws conclusions for them, leaving the majority saying that the teacher draws conclusions sometimes (40.7%) and almost always (39.4%). However, there is an increasing trend of students who claim that they draw their own conclusions almost always (52.2%), justify their conclusions (56.3%) and connect conclusions to scientific knowledge (61.4%). The latter signals student involvement in the process of drawing conclusions.

Collecting and analysing data

■ Never   ▧ Sometimes   ▨ Almost always

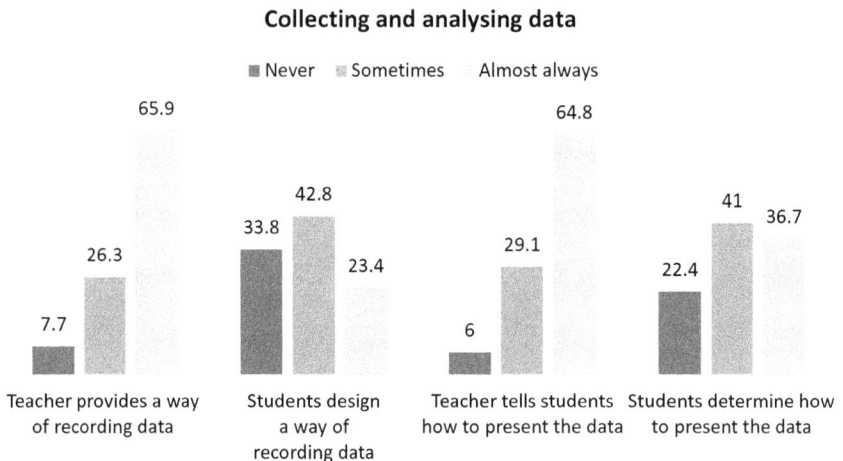

Figure 5.5 Teacher providing a way of recording data

Drawing conclusions

■ Never ■ Sometimes Almost always

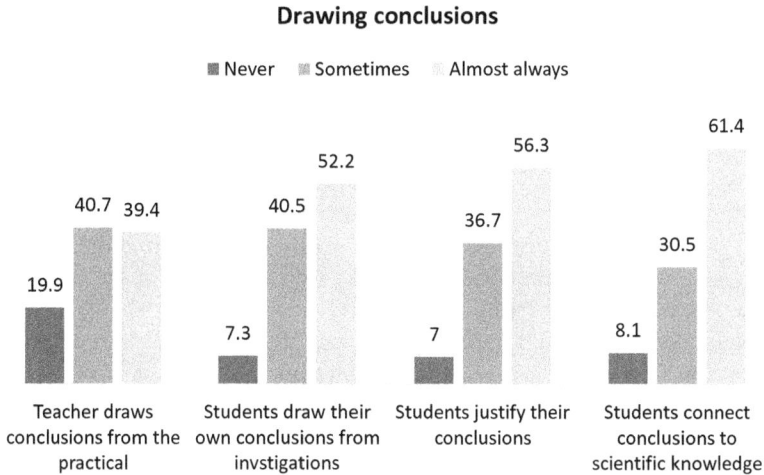

Figure 5.6 Drawing conclusions

Across the grades, the majority of students in all grades said that the teacher almost always draws conclusions from the practical.

Using the framework of the pedagogical orientation, the findings from the students are indicative of an interplay of three of the four pedagogical orientations, although leaning more on the DD and DI orientations. For example, DD orientation is evident where 66.9% of the students said that the teacher provides questions to investigate, 85.1% said the teacher gives step-by-step instructions or procedure and design a way for recording (65.9%) and transforming data (64.8%). However, the majority of the students are involved in data collection almost always (57.1%) and sometimes (58.2) as 'hands-on' experiences. In as far as this is done while following a given set of procedures, this approach could be categorised as a DI orientation. 66.2% of the students (23.4% almost always, 42.8% sometimes), on the other hand, said that they are given opportunities to design their own way of recording data and presenting data (36.7% almost always, 41% sometimes), which borders on a GI orientation. The GI orientation is also evident in drawing conclusions where although only about 39.4% of the students said that the teacher almost always draws conclusions for them, slightly over half of the students said that they draw their own conclusions and increasingly they justify them and connect them to scientific knowledge. Across the grades, the orientation still tends to tilt towards DD and DI, with a minimal shift to GI and OI in the senior grades of Form 3 and Form 4.

b) Teacher understanding of the role of practical work

Teacher understanding of the role of practical work was sought through post-observation interviews.

Figure 5.7 suggests that the significant role of practicals, as seen from teacher perspectives, is to develop laboratory skills and techniques. The skills include: observation, handling apparatus, taking measurements, analysis etc. One teacher who conducted practical work with his learners on food tests commented that through this practical work, apart from students understanding how to carry out food tests, they have also learnt about the various laboratory skills as follows:

> ...in today's practical work the message was, I wanted them to be able to carry out the practical work correctly, but also to make what is a correct deduction, to maybe make correct observations and deductions, also, they learn practical skills... how to handle maybe some apparatus when the learners are doing practical OK? But also they learn observation skills, that is, they learn to observe something and then maybe that is relevant [inaudible] yeah that is some of the things that I can say maybe takes place in practical work. [T8]

The second frequently mentioned role of practical work is that it is used to verify scientific facts. In describing the role of practical work, one teacher used the word 'expose' to advance the argument that practical work verifies theory the students learnt in class. He said it 'exposes' learners to real-life situations which eventually, he claims, fosters students' understanding of the concept. Other teachers mentioned that practical work helps students to remember. Yet one teacher mentioned that practical work brings enjoyment to the learners:

> Yeah, practical work, also helps learners to enjoy the lesson since they are able to do or since they are able to see what the teacher is talking about. They are doing; they are able to see the results, they are enjoying the lessons. [T6]

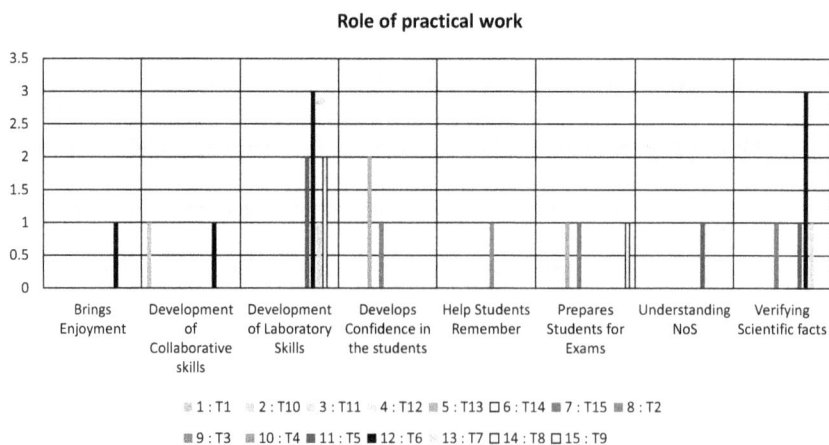

Figure 5.7 Teachers' views concerning the role of practical work

Other teachers mentioned that practical work helps in the development of collaborative skills. Three teachers also mentioned that practical work also prepares the learners for examinations. In the interviews, it was clear that in their professional practice they strive to do practical work with the learners as it helps the students to enhance their confidence regarding the handling of apparatus, a skill which is critical when it comes to examinations. For instance, T15 explained how practical work enhances students' confidence, which in turn helps them during examinations:

> Apart from that, it is important to do practical work because as a person maybe you can add confidence to the kids so since our education here is examination oriented, once they come across the questions on that particular practical in an exam, they can simply do it. So that's why they are supposed to do the practical so that they can have that confidence that they can handle the apparatus on their own. [T15]

One teacher linked practical work to the NoS. One of the tenets of the NoS is that science relies on evidence to support hypotheses or theories as he describes below:

> In fact, as the role of practical work is to verify as I have said but again it has also to give the students the mind of saying whatever people have said scientifically, we need to prove whether it is true or not true. So they are trying to prove, which can also come up with the critical thinking in the learners etc. Yeah they might have that to say... if this is true then they can also look at other things which are outside of the already stated concepts which... [T9]

The above comment suggests that the teacher is aware of the role practical work plays in enhancing the NoS.

Most of the teachers' views about the role of practical work link to the development of laboratory skills and techniques, verifying theory, giving a feel for phenomena and enhancing student motivation. Only one teacher linked practical work to the NoS and therefore the transferrable cognitive skills of critical thinking. What is essential from the teachers' perspective is that students acquire the necessary laboratory skills for them to verify the theory and pass examinations. This view might have influenced teacher pedagogical practice as most provided questions, procedures and a way of recording and analysing the data so that the role of the students was limited, for the most part, to handling apparatus and collecting data. Minimal attention was paid to enabling the students to behave like real scientists where they would explore questions whose answers they did not know and find solutions.

When asked specifically on how they would approach the teaching of practical work in order to achieve the purported roles of practical work, most teachers highlighted their roles which included giving instructions, provision of materials, guiding learners, facilitation and supervision.

...to make it more effective to the students, the practical work should be done based on the learner-centred approach. By saying learner-centred approach, give the students instructions to do and also give them some leading questions from each and every activity so that they should be able to discuss the question after doing the activity. What do you think is this, and what do you think is this? So from that, we are giving more opportunity to the learners to explore more ideas, more concepts unlike a teacher getting involved much in the activities – just as giving everything to them, they will not learn much so practical work should actually be based on a learner-centred approach. [T10]

This didactic orientation permeated the other roles that teachers mentioned such as facilitation, supervision and guiding learners. For example, when asked what was meant by facilitation, teachers said that facilitation may include 'helping the learners do the right thing' by 'giving them directions' (T6). Clearly, the facilitation is towards following set procedures so that the students find the 'right' answers. The same understanding was true under supervision:

...the teacher should be able to supervise here and there where students have struggles. I was supervising them in terms of looking at how they measure the pieces because sometimes I could see students struggling to measure length by width and height so I was trying to make sure that they should have accurate measurements. So I was trying to see how they are doing coz if they are just idle there you say everything is just OK; no, you are supposed to go around. [T11]

Other teachers also pointed out that one of the roles of the teacher during practical work is to consolidate the various conclusions students have come up with, as T1 explains below:

OK, not only facilitating but then you can also help the learners in doing the experiment if there are challenges right you can help them, maybe for example if you look at the conclusion like in my case, like if I ask the learners what is the conclusion you know they will be giving an answer but not maybe really exact. So you come in like to consolidate their responses. [T1]

According to T2, in the process of conducting practical work, students might make mistakes, especially during the conclusion. Hence, he explains it is the duty of the teacher to make a consolidation:

Thereafter students should be able to give whatever they have observed, or whatever they have you know experimented in practical. Then the teacher should just come to consolidate here and there to polish up whatever...

because the students can make mistakes so you just give clear, clear it up to them so that they should be on a good track. [T2]

These comments are illustrative of how much teachers struggle to give autonomy to the learners when they are conducting practical work. Not only are the teachers interested in following what the learners are doing on their own, but they are also interested in the extent to which learners are coming up with the desired conclusions.

Using the pedagogical orientation framework, what is clear is that the teachers have a DD orientation towards practical work. Firstly, by giving students procedures on how to conduct practical work, it suggests that the teacher has a high level of control on how the practical work or activity ought to be conducted. Secondly, the fact that the teacher gives students leading questions to interrogate the data or their findings implies that the teacher wants the students to come up with conclusions that are predetermined. These tendencies are typical characteristics of DI orientations, although they may have some GI tendencies. This was reflected in the teacher's practice as the teacher could be seen guiding the learners throughout the practical work.

Discussion

Teachers using the Direct Didactic orientation

The study used a framework of pedagogical orientation that had four categories: DD, DI, GI and OI. These types of orientation were explored using lesson observation and a student questionnaire with 18 items that described the process of inquiry in practical work with five categories: a) asking questions/framing research questions, b) designing investigations, c) conducting investigations, d) collecting data and e) drawing conclusions. Data from both observation and students' questionnaires show that teachers adopted a DD orientation in that teachers gave students questions to investigate and provided step-by-step instructions on how to conduct the procedure, and how to record and present data. It is only in conducting the procedure that students were actively involved and to a lesser extent in drawing conclusions. It may thus be summarised that teachers exhibited a DD orientation in the part of practical work involved in asking questions, designing procedures, and recording and presenting data. On the other hand, they allowed students to carry out the procedures, an aspect of DI orientation; and to a limited extent, draw conclusions which is part of GI. Similarly, students' experience, as shown by responses in the questionnaire, show these three orientations.

It is thus clear that teachers see practical work as involving hands-on experience for students, but they retain the control of the minds-on experiences, especially the formulation of the questions and design of the experience. However, they allow students, to a limited extent, to engage in drawing conclusions, justifying them and making links with theory. Some teachers were still drawing conclusions for the students, showing more control. On the aspect of DI, conducting procedures, all teachers allowed students to do the procedures and gave them considerable time. It would seem that this is an aspect that embodies practical work in the minds of their teachers. Teachers ensured that adequate materials, time and instructions or step-by-step procedure were available for this. Finally, the DD orientation was shown in the first two aspects of inquiry: asking questions and designing procedures. These are the thinking part of the practical work, and it may be that teachers did not see them as practical work or they were not confident that their students would do them. This suggests that teachers did not see these as part of practical work and therefore just provided them to the students. They did not see the need for students to develop these skills. In this regard, they may have not correctly interpreted the curriculum which spells out these skills when describing investigations. It is also possible that the teachers may not have experienced these aspects at school or during teacher training. This results in teachers having undeveloped PCK (Cochran, DeRuiter, & King, 1993; Magnusson et al., 1999) for handling the conduction of practical work in schools.

Teacher pedagogical orientations shaped by their understanding of the roles of practical work and how it should be taught

Teachers mentioned five roles of practical work in developing laboratory skills and techniques, such as: observation, handling apparatus, taking measurements and analysis; to verify theory; to pass the practical examination; to aid memory; and for enjoyment, which are similar to the roles identified by Abrahams (2011). Only one teacher mentioned the development of the NoS, which has a bearing on enabling students to behave like a 'scientist for a day', undergoing all the reasoning and thinking done by scientists. The observed teachers' views of the role of practical work somehow had a bearing on how the teachers taught practical work. A majority of the observed teachers leaned more towards giving autonomy to students to conduct procedures while retaining teacher control in the formulation of questions and design of investigations. It is possible that for some teachers, it was the hands-on experience which was important in practical work as the other skills can be developed in theory lessons. However, it is also possible that given the time constraints in teaching the long syllabuses, some teachers thought that it works equally well for students to undertake the practical activities in response to a question that was already provided. A departure from this is the finding which has shown that some of the teachers let students draw their own conclusions. This represents a shift towards GI. Almost all the teachers were of the view that students need to be guided throughout practical work by giving them instruction on how to conduct practical work, directing them to draw the conclusions which the teacher wants them to achieve and consolidating what is learnt. These views emanate from a didactic orientation, which, to some extent, has a low view of the students' ability or has prescribed low-level learning outcomes. Teachers lack confidence in their students' ability to be independent learners.

This finding suggests that for the practice of practical work to be changed, the teachers have to understand the varied roles of practical work. This has a bearing on how the teachers are taught during both pre-service and in-service teacher education. There is a need to begin the change in the practice of practical work in pre-service teacher education. What is observed is that in the teacher training institutions, students are given a laboratory manual with already planned experiments and data collection procedures and the students simply follow. This limits their understanding of practical work to simply collecting data and drawing conclusions.

Conclusion

In conclusion, the findings of the study seem to suggest that the practice of teaching practical lessons in Malawi, for most teachers, leans more towards the DI pedagogy, with a few teachers displaying GI pedagogy. While there could be a number of factors contributing to this practice, including contextual factors, the findings show that the teachers' perspectives and understanding of practical work also had a bearing on the practices. The teachers' perspective showed that, for them, the role of practical work is for acquisition of laboratory skills, verification of theory and to aid memory, and is taught using mostly didactic approaches (direct for minds-on and interactive for hands-on). In this regard, teachers did not appear to have a notion of practical work as involving designing and undertaking investigations (save for one teacher). It is possible that this conceptualisation arose, in part, from their teacher education experience (both science and science education courses), which emphasise use of a laboratory manual with pre-designed experiments. It is possible that if teacher training involves the students in designing their own investigations and undertaking them, some teachers could implement practical work in a similar way. However, such an assertion will need to be substantiated with research evidence.

Notes

1 There are three types of public secondary schools in Malawi: grant-aided, conventional and community day. The latter are owned by communities and often have limited infrastructure for teaching science while grant-aided schools are owned by churches and are well equipped. Conventional schools represent an average secondary school with a basic minimum of n=infrastructure and consumables for practical work.
2 The school should have a basic minimum of infrastructure.

References

Abrahams, I. (2009). Does practical work really motivate? A study of the affective value of practical work in secondary school science. *International Journal of Science Education*, 31(17), 2335–2353. doi:10.1080/09500690802342836

Abrahams, I. (2011). *Practical work in secondary science: A minds-on approach.* New York: Continuum.

Abrahams, I., & Millar, R. (2008). Does practical work really work? A study of the effectiveness of practical work as a teaching and learning method in school science. *International Journal of Science Education,* 30(14), 1945–1969. doi:10.1080/ 09500690701749305

Abrahams, I., & Reiss, M. J. (2012). Practical work: Its effectiveness in primary and secondary schools in England. *Journal of Research in Science Teaching,* 49(8), 1035–1055. doi:10.1002/tea.21036

Campbell, T., Abd-Hamid, N. H., & Chapman, H. (2010). Development of instruments to assess teacher and student perceptions of inquiry experiences in science classrooms. *Journal of Science Teacher Education,* 21(1), 13–30. https://doi.org/10.1007/s10972-009-9151-x

Cerini, B., Murray, I., & Reiss, M. (2003). *Student review of the science curriculum: Major findings.* London: Planet Science/Institute of Education, University of London/Science Museum. Retrieved July 31, 2015 from www.planet-science.com/sciteach/review

Cobern, W. W., Schuster, D., Adams, B., Applegate, B., Skjold, B., Undreiu, A., … Gobert, J. D. (2010). Experimental comparison of inquiry and direct instruction in science. *Research in Science & Technological Education,* 28(1), 81–96.

Cochran, K. F., DeRuiter, J. A., & King, R. A. (1993). Pedagogical content knowing: An integrative model for teacher preparation. *Journal of Teacher Education,* 44(4), 263–272.

Cohen, L., Manion, L., & Morrison, K. (2007). *Research methods in education* (6th ed). New York: Routledge.

Government of Malawi. (2013a). *Biology curriculum.* Malawi Institute of Education.

Government of Malawi. (2013b). *Chemistry curriculum.* Malawi Institute of Education.

Government of Malawi. (2013c). *Physics syllabus.* Malawi Institute of Education.

Hewson, M. G., & Hewson, P. W. (1983). Effect of instruction using students' prior knowledge and conceptual change strategies on science learning. *Journal of Research in Science Teaching,* 20(8), 731–743. doi:10.1002/tea.3660200804

Hodson, D. (1991). Practical work in science: Time for a reappraisal. *Studies in Science Education,* 19(1), 175–184. doi:10.1080/03057269108559998

Law, N. (2009). Mathematics and science teachers' pedagogical orientations and their use of ICT in teaching. *Education and Information Technologies,* 14(4), 309. doi:10.1007/s10639-009-9094-z

Lunetta, V., Hofstein, A., & Clough, M. P. (2007). Teaching and learning in the school science laboratory: An analysis of research, theory, and practice in S. K. Abell & N. G. Lederman (Eds.), *Handbook of research on science education* (pp. 393–441). Mahwah, NJ: Lawrence Erlbaum Associates.

Magnusson, S., Krajcik, J., & Borko, H. (1999). Nature, sources, and development of pedagogical content knowledge for science teaching. In J. Gess-Newsome & N. G. Lederman (Eds.), *Examining pedagogical content knowledge: The construct and its implications for science education* (pp. 95–132). Dordrecht: Kluwer Academic.

Millar, R. (2004). The role of practical work in the teaching and learning of science. A paper prepared for the committee High School Science Laboratories: Role and Vision, National Academy of Sciences, Washington, DC. Retrieved August 30, 2019 from https://sites.na tionalacademies.org/cs/groups/dbassesite/documents/webpage/dbasse_073330.pdf

Ramnarain, U.Nampota, D., & Schuster, D. (2016). Spectrum of pedagogical orientations of Malawian and South African physical science teachers towards inquiry. *African Journal of Research in Mathematics, Science and Technology Education,* 20(2), 119–130.

Science Community Representing Education (SCORE, 2008). A research report: *Practical work in science: A report and proposal for a strategic framework.* London: Gatsby

Technical Education Projects. Retrieved August 28, 2019 from https://score-educa
tion.org/media/3668/report.pdf

Solomon, J. (1980). *Teaching children in the laboratory*. London: Croom Helm.

Wellington, J. (Ed.). (1998). *Practical work in school science: Which way now?*London:
Routledge.

Woolnough, B., & Allsop, T. (1985). *Practical work in science*. Cambridge: Cambridge
University Press.

Appendix

Table A.1 Lesson observation schedule

A. Asking questions/framing research questions																
A1. Teacher gives students questions to investigate																
A2. Teacher asks students to formulate questions which can be answered by investigations																
A3. Teacher discusses student research questions to determine the direction and focus of the laboratory lesson																
A4. Teacher asks students to refine their questions so that they can be answered by investigations																
B. Designing investigations																
B1. Teacher gives step-by-step instructions before students conduct investigations																
B2. Students engage in a critical assessment of the procedures given by the teacher																
B3. Students design their own procedures for investigations																
C. Conducting investigations																
C1. Students conducting procedures of an investigation																
C2. Teacher conducting the investigation without involvement of the students																
C3. Teacher conducts investigations with students' involvement																
D. Collecting data, analysis and presentation																

D1. Teacher provides a way for recording data															
D2. Students design a way of recording data															
D3. Students determining how to present the data															
D4. Teacher tells students how to present the data															
E. Drawing conclusions															
E1. Students discuss a variety of ways of interpreting evidence when making conclusions															
E2. Students state their own conclusions for investigations															
E3. Students connecting conclusions to scientific knowledge															
E4. Students justifying their conclusions															

Teacher interview schedule

1   Did the lesson go according to plan?
2   What changes did you make and why?
3   What do you think students learnt through this practical? What else could they have learnt?
4   If you were to repeat, what would you do differently?
5   In the lesson observed, it was noted that you. Why?
6   To what extent are learners given opportunity to ask questions, design experiments, conduct investigations, collect data, analyse data, and draw conclusions?
7   What forms of practical work do you normally do in class?
8   What are the aims of practical work?
9   How do you match aims and type of practical work?

Table A.2 Student questionnaire

	Never	Sometimes	Almost always
A. Asking questions/framing research questions			
A1. Students formulate questions which can be answered by investigations			
A2. Student research questions are used to determine the direction and focus of the laboratory lesson			

	Never	Sometimes	Almost always
A3. Time is devoted to refining student questions so that they can be answered by investigations			
B. Designing investigations			
B1. Students are given step-by-step instructions before they conduct investigations			
B2. Students design their own procedures for investigations			
B3. Students engage in the critical assessment of the procedures that are employed when they conduct investigations			
B4. Students justify the appropriateness of the procedures that are employed when they conduct investigations			
C. Conducting investigations			
C1. Students conduct procedures of the investigation			
C2. The investigation is conducted by the teacher without involvement of the students			
C3. Students actively participate in investigations as they are conducted (whole class, demonstration, group)			
D. Collecting data			
D1. Students determine which data to collect			
D2. Students determine how to present the data			
D3. Students take detailed notes during each investigation along with other data they collect			
D4. Students decide when data should be collected in an investigation			
E. Drawing conclusions			
E1. Students consider a variety of ways of interpreting evidence when making conclusions			
E2. Students develop their own conclusions for investigations			
E3. Students connect conclusions to scientific knowledge			
E4. Students justify their conclusions			

Adapted from Campbell, Abd-Hamid, and Chapman (2010)

Chapter 6

Enactment of practical work in Kenyan secondary schools

Findings in a narrative inquiry

Josephat M. Miheso

Introduction

Science educators and teachers agree that practical work is indispensable to the understanding of science (Carrack, Onder, & Dikmenli, 2007). Science education aims at helping learners to gain an understanding of the established body of scientific knowledge as is appropriate to their needs, interests and capacities, and the understanding of the methods by which this knowledge has been gained (Duschl, Schweingruber & Shouse, 2007). Duschl et al. (2007) perceive science as both a body of knowledge that represents current understanding of the natural systems and as a process whereby this body of knowledge is being continually extended, refined and revised. This understanding is consistent with Schwab (1964), who identified two types of subject matter knowledge (SMK): substantive knowledge, which includes the organization of concepts, facts, principles, and theories, and syntactic knowledge that comprises the rules of evidence and proof used in making claims about new knowledge in the subject (Bullock, 2009). According to the National Research Council (2012), science education cultivates students' habits of mind, develops their capacity to engage in scientific inquiry and teaches them how to reason in a scientific way. As a scientific inquiry process, the teaching of practical work therefore follows the basic principle of learning by doing, where learners are given an opportunity to actively participate in the learning process, by engaging in concrete and authentic experiences. This mode of instruction involves first-hand experiences that permit learners to participate in learning as a way of thinking and investigation. Learners' involvement in scientific investigations leads to simultaneous development of the requisite procedural skills in science as well as a deeper understanding of scientific concepts, laws, and theories (Mattheis & Nakayama, 1988).

Practical work in science education can be used to promote many learning outcomes amongst which are helping learners such as helping learners to develop important practical skills, and understanding the process of scientific investigation that often leads to stimulation of their continued interest in learning science (National Research Council, 2012). Practical work can therefore be used as a platform for helping learners to better understand scientific concepts and principles (Freedman,

2002), by deliberately providing them with the opportunity to engage in concrete and authentic practical learning experiences.

This chapter adopts a narrative inquiry approach in reviewing the unfolding of practical work in Kenyan schools. According to Riessman (2008), narrative inquiry connects events in a sequence that is consequential for later action and for meanings that the narrator wants listeners to take away from the story. In this chapter, important aspects regarding the enactment of practical work in Kenyan secondary schools are organized and discussed in six sections. The next section provides an introduction to the role of practical work in science education. This is followed by a discussion of the approaches used for teaching practical work. The chapter continues with a review of selected literature relevant to practical work instruction in Kenya, which is followed by a reflection on the performance of learners' practical work summative evaluation in Kenya over a period of four years (2013 to 2017). The nature of science teacher preparation in Kenya is then discussed, and the chapter concludes with implications and recommendations for practical work instruction in Kenya.

Practical work and science education

The role of practical work in science education has been detailed by some researchers (e.g. Lazarowitz & Tamir, 1994; Tamir & Lunetta, 1981). According to Lazarowitz and Tamir (1994), practical work can be used to induce scientific perceptions, develop problem-solving skills and improve conceptual understanding. Accordingly, practical work provides learners with the opportunity to experience science using scientific research procedures (Hofstein, 2004).

The observations made during practical work and the results obtained help to systematically develop the science process skills necessary for the world of work (Manjit, Ramesh, & Selvanathan, 2003). In addition, abstract ideas can be concretized and naive, neonate and scientifically primitive ideas challenged (Musasia et al., 2012). In this way, the desire and eagerness to know more about what the subject can offer is developed.

The argument above implies that science teachers have to design instructional methods that stimulate learners' thinking processes. They (teachers) need to subsequently make explicit how each aspect of practical inquiry activity reflects scientific investigation and how this relates to learners' daily life experiences. The reality on the ground however is that most of the practical work conducted in many developing countries is ill conceived, and not effective in getting learners to use the intended scientific ideas to guide their actions and reflect upon the data (Abrahams & Millar, 2008). Accordingly, a great deal of laboratory work that takes place in schools is aimless, trivial and badly planned, moreover, practical lessons are often too short for learners to complete the planned practical work tasks in time (Machina, 2012), Furthermore, the teaching and learning resources/equipment for practical work are limited in many schools.

Approaches to teaching science practical work

According to the National Research Council 2006 report on high school science enhancement of student learning, teachers should be able to:

1   specify the learning outcomes in measurable terms;
2   thoughtfully sequence laboratory work with other types of instruction;
3   integrate the process of finding out with learning the content under study; and
4   incorporate student reflection and discussion through laboratory work.

Practical work in science can thus be used to achieve many different learning outcomes that include verifying concepts previously discussed in classrooms, development of particular manipulative skills needed in subsequent practical work and facilitating attainment of various concepts. The desired practical outcomes therefore specify the type of practical work enacted during classroom instruction. According to Prince and Felder (2006), the two different approaches for conducting scientific learning are deductive and inductive reasoning.

Deductive or verification practical work

Prince and Felder (2006) contend that deductive reasoning involves testing a theory by collecting and examining empirical evidence to see if the theory is true. The purpose of deductive practical work is to subsequently confirm concepts, principles and laws previously discussed during classroom discussion. Generally, most science teachers present major scientific concepts, ideas and principles in the classroom through lecture and discussion methods, followed by laboratory experiments to illustrate examples, as they try to verify various attributes and relationships. Many of the laws in science can be illustrated in the laboratory by the deductive approach method. Verification practical activities have the advantage of providing learners with an idea of what they expect to find out in advance about an abstract idea. This mode of practical learning helps to reinforce the subject matter content taught in classrooms.

Inductive practical work

Unlike deductive reasoning, inductive reasoning begins with specific observations or real examples of events, trends or social processes (Prince & Felder, 2006). Using this data, learners can then progress analytically to broader generalizations and theories that help explain the observed cases. Inductive practical work therefore provides learners with the opportunity to develop scientific concepts, principles and laws through first-hand experiences before discussing them in classrooms. This mode of practical work exposes learners to experiences that enable them to search for patterns and identify relationships from experimental data. Ideas and theories are thereafter discussed in class under the teacher's guidance and applications of the concepts provided to reinforce learning. Inductive reasoning as an approach to teaching practical work should thus be encouraged because meaningful practical work is often embedded in the discussion of ideas emanating from observations and findings against scientific principles and theories. This mode of instruction is however dependent on inspirational and knowledgeable teachers

Science teaching in the Kenyan school curriculum

The government of Kenya recognizes the importance of science and mathematics in the realization of the country's Vision 2030, which aims at providing globally competitive quality education, training and research for development where all citizens have enhanced entrepreneurial, innovative and lifelong learning opportunities (A National Education Sector Plan [NESP 2013/14–2017/18]). This aspiration is reflected in the huge resources, both human and otherwise, that are channelled towards education. The percentage of the national GDP allocated to education out of the total national budget has been sustained at close to 16 per cent of GDP; a quarter of the total national budget has been spent on education for the last five years. Most of this funding is channelled towards enhancing the teaching and learning of science and mathematics (Migosi, 2017).

In Kenya, science education was prioritized after the country's independence in 1963, as formal education was largely not available before independence. Prior to independence, the system of education in Kenya was discriminatory, favouring a minority non-indigenous population. Consequently, only a small number of privileged indigenous learners attended school. Other than the discriminatory school education system inherited at the dawn of independence, the science teaching approaches in use at the time could best be described as mere 'cookbook' (Ogunniyi, 1986). The main aim of teaching science was to prepare learners for further studies in science-related courses and hopefully for them to be able to apply the learnt skills in their everyday activities. Learners were accordingly taught basic principles of science and practical skills through standard topics and experiments. This primary aim seems to have remained the same even after the change from the first post-independence education system (Republic of Kenya, 1964–65) to the current system of education, termed the 8-4-4 system. This is evident from the aims of the secondary education level, which require learners to be provided with the opportunity to build a firm foundation for further education and training, acquire necessary knowledge, skills and attitudes for the development of the self and the nation, and develop the ability for enquiry, critical thinking and rational judgement among others (Kenya Institute of Education (KIE), 2002). The current system of education has been termed the 8-4-4, because learners attend (8) years of primary school, (4) years of secondary school, while those selected to join university education take at least (4) years to complete most of the undergraduate degree course programmes at any of the Kenyan universities (Wasanga & Somerset, 2013).

The 8-4-4 education system allows learners to study science as a compulsory subject using a common syllabus for the first 12 years of schooling. This phase of education comprises both the primary and secondary school levels of basic education. At the primary school level, science is taught as an integrated single subject, while at the secondary school level, science is divided into and taught as three distinct subject domains: Biology, Chemistry and Physics (Kenya Institute of Education (KIE), 2002). The teaching of science at secondary school is presented in the syllabi as practical subjects, where scientific concepts, principles and skills are expected to be developed through experimental investigations (KIE, 2002). However, fully fledged practical work is not a common feature in many Kenyan schools, including even some of the schools with sufficient teaching and learning resources. According to Sifuna and Kaime (2007), many teachers in Kenya carry out very few science demonstrations and almost no classroom experiments. The authors argue that although small group practical work is done, follow-up discussions on the purpose of the exercises performed are usually counter-productive. In addition, the teaching and learning materials, resources and equipment for practical work are limited in many schools. Learners therefore frequently follow fixed recipe-type programmes of experimental manipulations and observations set by the teacher. Orado (2009) observes that although learners are involved in a variety of practical activities during their practical work, they only develop basic scientific process skills, leaving out key integrated scientific skills such as experimental design and hypothesis formulation.

Secondary education is a strong driver of skills for learning, employability and decent work in Kenya. There is an urgent need to therefore align secondary education to the skills needed in today's workforce and the social values envisioned for the country. In response to this challenge, the country is currently reforming its education, from the elitist, examination-oriented 8-4-4 system of education to a competency-based curriculum, with a focus on acquisition of competencies and nurturing learners' talents (Basic Education Curriculum Framework (BECF, 2019). However, there are still five years before the new curriculum is to be rolled out at secondary school level. The first class of graduates under the new system of education will be realized in 2028. This means there are many learners who will still graduate from high school under the current 8-4-4 curriculum between now and the full implementation of the Competency Based Curriculum (CBC), and the need for this discussion.

Teaching secondary school practical work in Kenya

In Kenya, secondary schools are classified into two main categories: public and private. The public schools, which number approximately 7,325, form the majority of schools, while there are about 921 private schools in the whole country (MoEST, 2015). For the purposes of admitting learners into secondary schools, the public schools are further categorized into four clusters. It is worth noting at this point that Kenya as a country, is sub-divided into 47 county administrative units; each county is further sub-divided into smaller adminis-trative sub-units called sub-counties or districts.

The first category of schools comprises national schools; there are about 103 in number and they account for approximately 5 per cent of the total second-ary school population. This category of schools admits the best-performing candidates across the country, based on their performance in the Kenya Certi-ficate of Primary Education (KCPE) summative evaluation examination, which is conducted at the end of the primary school level of education. National schools have well-established infrastructure, in terms of well-equipped science laboratories, as well as adequate supply of teaching and learning resources for practical work. The second cluster of schools is referred to as extra-county schools. These are high-performing boarding secondary schools, and there are about 331 in number. The extra-county schools admit 40 per cent of their learners from across the country while the rest come from the host county. The third cluster of schools comprises the county boarding secondary schools. This are average-performing schools, the majority of which are on average well-resourced, in terms of teaching/learning facilities/resources for practical instruction. However, some of the schools have well-established infrastructure, with sufficient supply of teaching/learning facilities and resources for practical work. The county schools admit learners from the host county only.

The last cluster of schools comprises the day schools or boarding schools with a day wing, referred to as the sub-county schools. Sub-county schools admit their learners from the host sub-counties or districts only which, as alluded to above, are county sub-units. This cluster of schools are about 5,699 in number and form the majority of secondary schools in Kenya, with a student popula-tion of about 1.3 million (MoEST, 2015). Most of the sub-county schools lack the basic teaching/learning facilities/resources for practical work. The majority of the schools have one or two science laboratories that are often ill-equipped for effective practical work instruction. Some schools in this category do not have any science laboratory/room for performing practical work. This situation is further aggravated by high learner enrolments, thus putting more strain on the little available teaching/learning resources.

Despite differences in facilities, leading to unequal provisioning of learning experience, learners from the four clusters of schools are nevertheless expected to compete equally for the same placements at the state-run universities and tertiary learning institutions. This is despite the fact that national schools and extra-county schools expose their learners to practical work quite early, at the beginning of their secondary school course programme. On the contrary, some of the sub-county schools expose their learners to practical work experiences towards the end of the four-year Kenya Certificate Secondary Education

(KCSE) course programme. Most learners in this category of schools often come into contact and get a feel for science practical work for the first time during their final practical summative assessment examinations, discussed below.

Many sub-county schools across the country therefore hardly prepare their learners for practical work because of a scarcity of science learning equipment/ resources (Mabatuk, 2014). The dearth of practical work in some of the schools implies that some of the secondary school learners do not attain mastery of the requisite basic science practical skills necessary for their daily life experiences by the end of their basic education level. This level of education is unfortunately terminal for most of the learners.

The high input of resources being channelled towards enhancing the teaching and learning of science and mathematics in Kenya does not seem to be bearing fruit. This position is reflected in the learners' performance in their summative practical work assessment conducted at the end of the secondary school course programme, outlined in the next section.

Performance in science practical work in Kenyan secondary schools

Flowing from the above discussion, the objective of teaching science could be said to fall into two main categories: an understanding of concepts and acquisition of a range of manipulative and process skills. Practical work emphasizes the latter aspects. Manipulative skills are associated with motor skills, and relate individual cognitive function with corresponding physical movement (Kempa, 1986). In science, manipulative skills emphasize the usage and handling of scientific apparatus and chemical substances during scientific investigation in the laboratory. In addition, students are exposed to the proper technique for using, cleaning and storing scientific equipment safely (Fadzil & Saat, 2014). On the other hand, process skills are more cognitive in nature.

Moreover, good laboratory techniques are essential for conducting successful practical work and collecting accurate data. Skills such as cutting glass and setting up apparatus and equipment for experimental work are essential during practical science instruction. According to Hegarty-Hezel (1990), science educators place little emphasis on developing proficiency in laboratory skills and techniques yet both teachers and learners require gradual development of these basic science process skills for effective participation in the teaching and learning process. Manipulative skills are generally given the least amount of attention in the course of academic instruction (Trowbridge, Bybee & Powell, 2000). Psychomotor and mental skills practice is also necessary for improving learners' abilities to make accurate and precise experimental investigations. Physical practice with laboratory equipment provides concrete experience with the apparatus and procedures during class time and discussion sessions, which is then followed by mental practice of the skills and procedures to ensure better conceptualization of the subject matter.

Science process skills can be grouped under two main categories: basic science process skills and integrated science process skills (Ergül et al., 2011). Basic process skills provide an intellectual grounding in practical work, especially in elementary grades (Rambuda & Fraser, 2004). These are the skills emphasized at the secondary school level of education in Kenya, which learners are required to master before acquiring the advanced integrated science process skills. Basic process skills are briefly outlined below.

- Observation comprises using one or more of the five senses to gather information about objects and/or events. When making observations, learners often look for similarities and differences as they try to compare and contrast their observations.
- Predicting involves stating the expected outcomes of a future event based on a pattern of evidence. Predictions are based on prior knowledge gained through experiences or data collected. Patterns or general trends in the information collected allow learners to make predictions about something beyond actual observations or measurements.
- Measuring involves using quantitative observations, standardized measuring tools or non-standardized objects to collect and record scientific data.
- Inferring encompasses explaining observations and drawing conclusions based on information or knowledge of cause and effect or past experiences. One is able to recognize patterns and identify relationships among data and make generalizations. Applications of concepts are henceforth provided to reinforce learning.
- Classifying begins with observing similarities and differences among objects and/or events in order to categorize them according to a predetermined set of properties or schemes.
- Communication is associated with giving or exchanging information. It includes using words and/or graphic symbols to describe an action, an object or an event either orally or in writing.

The mastery of both manipulative and process skills in science practical work enables learners to conceptualize the content they do know at a much deeper level of understanding, thus equipping them with skills for acquiring content knowledge in the future (Sevilay, 2011).

The second set of skills comprise the integrated science process skills. These are the immediate skills used in problem solving or in doing science experiments (Rambuda & Fraser, 2004). Integrated skills include skills for controlling variables, formulating hypotheses, interpreting data, experimenting and project work (Ongowo & Indoshi, 2013). Integrated skills are therefore critical for data collection, performing various scientific investigations and producing evidence to answer scientific questions. These skills are structured on basic skills, and are mostly emphasized at the tertiary level of education. Some of these skills are also applied in project work at secondary schools.

In Kenya, practical skills are assessed separately from the theoretical knowledge acquired by learners at the end of the secondary school course programme in the three science subjects named above. It is mandatory for all candidates who sit the final summative examination, the Kenya Certificate of Secondary Education (KCSE), to select at least two of the three science subjects at form 2 (Grade 10) as part of their elective subjects of study during their secondary school level course programme (KIE, 2002). The Kenya National Examinations Council (KNEC), which is the government body mandated to conduct public examinations at basic and tertiary levels of education in Kenya, specifies that candidates are separately assessed in both theory and practical aspects in each of the three science subjects.

The content attributes and skills acquired by learners are tested by the Kenya National Examinations Council (KNEC, 2017) using two theory papers (1 & 2), and one practical paper 3 for each of the science subjects at the end of the four-year secondary course programme (KIE, 2002). Paper 1 covers short answer compulsory test items (28–29) that cut across the entire syllabus; it therefore addresses the aspect of paper validity. On the other hand, paper 2 comprises four compulsory long structured question items that consist of a series of several sub-questions or parts, requiring short answers, often on the same topic/theme or concept. The aim of the structured question items is to assess in-depth understanding of the subject matter knowledge in each of the three science subjects. The practical paper 3, which is comprised of two to three compulsory question items selected from any part of the syllabus, evaluates acquisition of both manipulative and basic process skills. The two theory papers (1 and 2) are both marked out of 80 marks, while paper 3 is marked out of 40 marks. The total raw mark, which is 200 marks, is then converted to a percentage score for each of the three subjects. It is however mandatory for a learner to pass the practical paper 3 in order to be awarded a pass in any of the three science subjects (KNEC, 2005).

According to the Kenya National Examinations Council (KNEC, 2017), some of the science topics prescribed in the syllabus are more frequently assessed than others. According to the report, the most commonly examined topics include the following:

Chemistry

- Simple classification of substances – separation techniques, use of apparatus and accurate recording of observations
- Salts – analysis of various types of salts in different states
- The mole – titration, molar solutions, calculations and use of standard solutions
- Organic chemistry – analysis of organic compounds qualitatively and recording of observations

- Acids, bases and salts – complex ions identification and salt analysis (qualitatively)
- Energy changes – quantitative determination of molar enthalpies (ΔH)
- Chemical kinetics – determination of rate of reaction

Biology

- Classification
- Microscopy
- Cell physiology
- Food test
- Ecology
- Growth and development
- Reception and response
- Support and movements in animals

Physics

- Measurements
- Forces-Hooke's law, equilibrium and centre of gravity
- Heat transfer – quantity of heat
- Light – lenses, refraction
- Electricity – cells
- Linear motion
- Gas law
- Electronics

The report further outlines the topics/sub-topics where candidates were found to consistently perform poorly in the previous examinations, which are as follows:

Chemistry

- Salts – analysis of various types of salts in different states
- The mole – titration, molar solutions, calculations and use of standard solutions
- Energy changes – quantitative determination of molar enthalpies (∇H)

Biology

- Classification
- Microscopy
- Support and movements in animals

Table 6.1 Candidates' overall performance in practical work for the years 2013 to 2017

Year	Paper	Candidature	Maximum score	Mean score	Standard deviation
2013	1. Biology	397,319	40	12.88	7.64
	2. Physics	119,819	40	22.85	7.98
	3. Chemistry	439,765	40	14.67	5.68
2014	1. Biology	432,977	40	20.82	8.39
	2. Physics	131,100	40	19.68	6.78
	3. Chemistry	476,582	40	17.57	6.19
2015	1. Biology	465,584	40	22.62	9.15
	2. Physics	139,100	40	22.71	7.62
	3. Chemistry	515,799	40	20.37	7.15
2016	1. Biology	509,982	40	10.99	6.76
	2. Physics	149,760	40	17.15	6.56
	3. Chemistry	566,836	40	13.63	6.31
2017	1. Biology	545,663	40	7.68	5.05
	2. Physics	160,182	40	19.33	8.33
	3. Chemistry	606,515	40	14.1	6.11

Physics

- Measurements
- Light – lenses, refraction
- Electricity – cells
- Linear motion

Table 6.1 provides a sample of learners' national performance statistics in the KCSE practical examinations in the three science subjects at the end of form 4 (Grade 12) between 2013 and 2017 (KNEC, 2017).

The overall learners' performance in practical work, shown in Table 6.1, reveals that generally the mean scores in the three subjects have remained consistently low, despite the increase in the number of candidates over the years. The performance in Biology practical indicates a massive decline in performance compared to the other subjects.

The implication from the findings above is that it might not be possible for Kenya to create the critical mass of a scientifically literate workforce that can meet the country's projected Vision 2030 of becoming a middle-income economy, unless urgent intervention measures in the teaching and learning of practical work in science are put in place.

Teacher preparation for science practical work

As alluded to earlier, science teaching in the Kenyan secondary school curriculum is experiential in nature. Learners are given an opportunity to develop science skills, concepts and principles through experimental investigations, demonstrations, fieldwork and excursions. However, the training of science teachers in school-type science practical work appears to be neglected in many teacher training programmes. According to Masingila and Gathumbi (2012), there are no school-type laboratories set aside to train teachers specifically for this purpose at universities.

The implication of the above posit is that science teachers are largely trained in theoretical content at the expense of developing inquiry practical skills. This could be one of the contributing factors to the poor teaching of practical work in secondary schools. In the same vein, Musasia, Abacha and Biyoyo (2012) argue that teaching secondary school science in Kenya is geared around memorization of basic concepts and their regurgitation in final examinations. One can therefore surmise that the practical work component does not receive the requisite attention during the pre-service teacher training programmes, hence the poor preparedness of science teachers for practical work in schools.

The other problem is the low enrolment and poor performance in science subjects, which is particularly noticeable among girls in Kenyan secondary schools (Ng'etich, 2014). For example, the low enrolment in upper secondary school physics has been linked to a shortage of inspirational and well-trained physics teachers, inadequate laboratory facilities and the accompanying limited exposure to practical instruction at junior secondary school level (Musasia, et al., 2012; Ng'etich, 2014).

In Kenya, a number of education intervention strategies have been put in place to ensure that the teaching and learning of science subjects remains effective at secondary school level. For instance, the government has institutionalized an In-service Education and Training (INSET) for serving science and mathematics teachers. The INSET programme, termed Strengthening of Mathematics and Science in Secondary Education (SMASSE), is aimed at upgrading mathematics and science teachers' skills for improved/enhanced classroom delivery of lessons. Studies on the impact of the SMASSE initiative on classroom performance in practical work (Mutisya, 2011; Orado, 2009) show marked improvement in the performance of mathematics and science in the KCSE examinations as well as a positive change in attitude towards mathematics and science learning among learners.

Conclusion

It is apparent that although practical work in science is important for good science teaching, its potential still remains to be fully exploited by science teachers in Kenya. The discussion further illuminates some of the critical factors that hinder learners' achievement of the requisite scientific practical skills to include lack of resources, inadequate laboratory facilities and resources for teaching, as well as the pedagogical approach used in science instruction. Naidoo (2007) argues that inadequate resources (both physical and human) for effective practical work are among the main causes of the deteriorating position of developing countries in the sciences.

There is a need for science teachers in developing countries, like Kenya, to therefore strategically plan how to utilize their inevitably limited resources in order to help learners connect scientific concepts to concrete experiences and improve the quality of science learning in secondary schools, in order to effectively participate in the emerging technologically knowledge-based economy. The discussion further points to the long-term value of practical work in science education, and hence the need to underscore the importance of practical work in pre-service teacher training programmes.

References

Abrahams, I., & Millar, R. (2008). Does practical work really work? A study of the effectiveness of practical work as a teaching and learning method in school science. *International Journal of Science Education*, 30(14), 1945–1969.

Bullock, S. M. (2009). Learning to think like a teacher educator: Making the substantive and syntactic structures of teaching explicit through self-study. *Teachers and Teaching: Theory and Practice*, 15(2), 291–304.

Carrack, O., Onder, K. & Dikmenli, M. (2007). Effect of the usage of laboratory method in primary school education for the achievement of the students' learning. *Asia-Pacific Forum on Science Learning and Teaching*, 8(2), Article 3. [Online]

Dewey, J. (1996). The theory of inquiry. In J. A. Boydson (Ed.), *John Dewey: The later works 1925–1953* (vol. 12, 1938). Carbondale, IL: Southern Illinois University Press.

Duschl, R. A., Schweingruber, H. A., & Shouse, A. W. (Eds.). (2007). *Taking science to school: Learning and teaching science in grades K-8*. Washington, DC: National Academies Press.

Ergül, R., Şımşeklı, Y., Çaliş, S., Özdılek, Z., Göçmençelebı, Ş., & Şanli, M. (2011). The effects of inquiry-based science teaching on elementary school students' science process skills and science attitudes. *Bulgarian Journal of Science & Education Policy*, 5(1).

Fadzil, H. M., & Saat, R. M. (2014). Enhancing STEM education during school transition: Bridging the gap in science manipulative skills. *Eurasia Journal of Mathematics, Science & Technology Education*, 10(3), 209–218.

Freedman, M. P. (2002). The influence of laboratory instruction on science achievement and attitude towards science across gender differences. *Journal of Women and Minorities in Science and Engineering*, 8(2), 191–199.

Hegarty-Hazel, E. (1990). *The student laboratory and the science curriculum*. London: Routledge.

Hofstein, A. (2004). The laboratory in chemistry education: Thirty years of experience with developments, implementation, and research. *Chemistry education research and practice*, 5(3), 247–264.

Kempa, R. (1986). *Assessment in science*. Cambridge: Cambridge University Press.

Kenya Institute of Curriculum Development (KICD) (2019). Basic Education Curriculum Framework. Nairobi, Kenya: KICD.

Kenya Institute of Education (KIE) (2002). *Secondary syllabus volume two*. Nairobi. Kenya: Institute of Education.

Kenya, Republic of (2007). *Kenya Vision 2030.* Nairobi: Ministry of Planning, National Development and Vision 2030.

KNEC (2017). *Kenya national examination council newsletters.* Nairobi: KNEC.

KNEC (2005). *Kenya Certificate of Secondary Education; Regulations and syllabuses, 2006–2007.* Nairobi: KNEC.

Lazarowitz, R., & Tamir, P. (1994). Research on using laboratory instruction in science. In D. L. Gabel (Ed.), *Handbook of research on science teaching* (pp. 94–130). New York: Macmillan.

Mabatuk, V. (2014). *KCSE candidates see Science practical apparatus for the first time hours to exams.* Retrieved from www.standardmedia.co.ke/article/2000138779/kcse-candida tes-see-science-practical-apparatus-hours-to-exams (accessed 20 October 2014).

Machina, M. J. (2012). *Prospective teachers' preparedness to facilitate chemistry instruction at secondary school level in Nairobi teaching practice zone – Kenya.* Unpublished master's thesis, Kenyatta University, Kenya.

Manjit, S. S., Ramesh, S., & Selvanathan, N. (2003). Using multimedia to minimize computational effort in engineering. *Proceedings of the Malaysian Scientific & Technology Congress (MSTC)* (pp. 811–815).

Masingila, J. O., & Gathumbi, A. W. (2012). A collaborative project to build capacity through quality teacher preparation. *Quality Education for Societal Transformation.* Nairobi, Kenya, July 20–22, 2011, 20, 69.

Mattheis, F. E., & Nakayama, G. (1988). *Effects of a laboratory-centered inquiry program on laboratory skills, science process skills, and understanding of science knowledge in middle grades students* (ERIC no. ED307148).

Migosi, J. (2017). *Secondary Education Quality Improvement Project (SEQIP) Vulnerable and Marginalized Groups Framework (VMGF).* Ministry of Education, State Department of Basic Education.

Ministry of Education Science and Technology (MoEST) (2015). A National Education Sector Plan (NESP 2013/14–2017/18). Nairobi: MoEST.

Musasia, A. M., Abacha, O. A., & Biyoyo, M. E. (2012). Effect of practical work in physics on girls' performance, attitude change and skills acquisition in the form two–form three secondary schools' transition in Kenya. *International Journal of Humanities and Social Science,* 2(23), 151–166.

Mutisya, B. P. (2010). *The impact of SMASSE project on the teaching methodology and learning in secondary schools in Kangundo/Matungulu district.* Post Graduate Diploma in Education Project, University of Nairobi.

Naidoo, R. (2007). Higher education as a global commodity: The perils and promises for developing countries. *The Observatory on Borderless Higher Education,* 1.

National Research Council (2006). *America's lab report: Investigations in high school science.* Washington, DC: National Academies Press.

National Research Council. (2012). *A framework for K-12 science education: Practices, cross-cutting concepts, and core ideas.* National Academies Press.

Ng'etich, J. K. (2014). *Factors influencing girls' low enrolment and poor performance in physics: The case of secondary schools in Nandi South District.* Unpublished master's thesis, Kenyatta University, Kenya.

Ogunniyi, M. B. (1986). Two decades of science education in Africa. *Science Education,* 70(2), 111–122.

Ongowo, R. O., & Indoshi, F. C. (2013). Science process skills in the Kenya Certificate of Secondary Education biology practical examinations. *Creative Education*, 4(*11*), 713–717.

Orado, G. (2009). *Factors influencing performance in chemistry practical work among secondary schools in Nairobi Province*. Unpublished M. Ed. thesis, Kenyatta University, Kenya.

Prince, M. J., & Felder, R. M. (2006). Inductive teaching and learning methods: Definitions, comparisons, and research bases. *Journal of Engineering Education, 95*(2), 123–138.

Rambuda, A. M., & Fraser, W. J. (2004). Perceptions of teachers of the application of science process skills in the teaching of Geography in secondary schools in the Free State province. *South African Journal of Education*, 24(1), 10–17.

Riessman, C. K. (2008). *Narrative methods for the human sciences*. Thousand Oaks, CA: SAGE Publications.

Schwab, J. (1964). The structure of the natural sciences. In G.W. Ford & L. Pugno (Eds.), *The structure of knowledge and the curriculum* (pp. 31–49). Chicago: Rand McNally.

Sevilay, K. (2011). Improving the science process skills: Ability of science student teachers using I diagrams. *Eurasia Journal of Physics & Chemistry Education*, 3, 26–38.

Sifuna, D. N., & Kaime, J. G. (2007). The effect of in-service education and training (INSET) programmes in mathematics and science on classroom interaction: A case study of primary and secondary schools in Kenya. *Africa Education Review*, 4(1), 104–126.

Tamir, P., & Lunetta, V. N. (1981). Inquiry related tasks in high school science laboratory handbooks. *Science Education*, 65, 477–484.

Trowbridge, L. W., Bybee, R. W., & Powell, J. C. (2000). *Teaching secondary school science*. Englewood Cliffs, NJ: Prentice Hall.

Wasanga, P., & Somerset, A. (2013). Examinations as an instrument for strengthening pedagogy: Lessons from three decades of experience in Kenya. *Assessment in Education: Principles, Policy & Practice*, 20(4), 385–406.

Chapter 7

Science practical work and its assessment in Ugandan secondary schools

Israel Kibirige

Introduction

Education in Uganda is highly prioritised for socio-economic development. This priority has borne dividends, and this was shown in the White Paper on Education that was commissioned by the Ugandan Government in 1992 (O'dama, 2013). Important improvements were noted such an improvement in the quality of teaching and learning through training more science teachers, the building of new science laboratories, and the refurbishment of old laboratories. The Ministry of Education and Sports (MoES) implemented the African Union Ministers of Education Goals for Education (1997–2005 and 2006–2015). The goals for 1997–2005 were equity and access to primary education, quality education, relevance and practical education, complementary learning modalities, and capacity building. A further commitment to improve the quality of education is underlined by the government of Uganda increasing the education budget to 16% of the national budget, and the creation of the National Curriculum Development Centre (NCDC, 2006) to plan, develop and evaluate secondary education curricula. Despite the success in implementing the African Union Ministers of Education Goals for Education in Uganda, such as improvement in youth literacy rates (O'dama, 2013), little is known about science practical work enactment and its assessment. Therefore, this chapter reports on empirical evidence of how practical work is enacted and assessed at two secondary schools from the Kampala District.

Ugandan secondary education comprises six years of education: four years for Ordinary Level (O-Level) (ages 12 to 16) and two years for Advanced Level (A-level) (ages 17 to 18). Learners at O-Level study 10 subjects to understand basic tenets in science, and at A-Level, the subjects are reduced to four and focus more intensely on the content than is the case for subjects at O-Level. The learners' interests and abilities are considered when choosing and allocating the four subjects at A-Level.

There are three categories of schools: low, middle and high-performing schools. The Department of Education and Sports categorises secondary schools as government or privately sponsored. These schools can be mixed-sex:

admitting boys and girls or single-sex: admitting boys or girls only. Regardless of the category, these schools use the same curriculum and as such the learners are assessed at their levels (O and A) in theory and practical work examinations administered by the Uganda National Examination Board (UNEB) for both.

A high stakes examination, the Uganda Certificate of Education (UCE), is taken at the end of the senior phase, and after two years of study, the Uganda Advanced Certificate of Education (UACE) examinations can be attempted. Since 1996, the National Assessment of Progress in Education (NAPE) has been conducted in primary Grades 3 and 6 and in 2008 NAPE was extended to secondary schools. There are also system-wide tests: practice examinations such as formal and informal assessments. Thus, Uganda's assessment system is complex and uses different forms of assessments at different academic levels.

The assessment of practical work constitutes three dimensions as stipulated by Stroupe (2015): 1) a conceptual dimension whereby learners are tested on ideas, laws and theories; 2) the development of explanations and solutions; and 3) evaluation and arguments about the reliability of the data collected. Teachers in Uganda use direct teaching to inculcate science content, and the content is applied by learners to perform practical work (Ssempala, 2017). At times, teachers' classroom practice is at odds with policy, especially on inquiry teaching (Altinyelken, 2010). Although curriculum policy stipulates the time allocated to practical work (NCDC, 2013), due to time constraints in covering the content teachers do not conduct practical work in some topics. The direct teaching approach results in learners being passive in class and the lessons being dominated by the teacher. However, it has been reported that it can improve learners' scientific thinking in two areas: reasoning abilities and domain-specific knowledge where teacher talk is interjected with learners' views and talk to break the passiveness that characterises direct teaching (van der Graaf, van de Sande, Gijsel & Segers, 2019).

Enactment and assessment of practical work in schools in Uganda

Practical work in secondary schools is specified to take place once a week for every science subject: Biology, Physics and Chemistry. The schools conduct practical work with the help of laboratory assistants, who are trained at various colleges in the country to work in secondary schools. In well-resourced schools, there are three laboratory assistants: one for each of Biology, Physics and Chemistry. In less-resourced schools, there are two laboratory assistants: one for Biology, and one for Physics and Chemistry.

There are three types of assessment for practical work suggested for schools. Firstly, *assessment of learning*, which uses evidence of student learning against the outcomes and standards. Sometimes this is referred to as summative assessment. It takes place during the stake practical examination. Secondly, *assessment for learning* is where teachers use evidence about learners' knowledge, understanding and skills to inform their teaching strategies. This is called formative assessment. This type of assessment takes place throughout the teaching and learning process. Thirdly, *assessment as learning* involves recognising learners' autonomy. Unfortunately, this type of assessment does not apply to secondary practical work. The process of assessment is complex, and it is contended that merely identifying the knowledge gap learners have to close in science content may not be adequate (Torrance, 2012). Assessment involves exploring and exploiting the gaps between teacher and student and between learners' present knowledge and developing pedagogic action so that the learners can understand. This level of complexity is seemingly not reflected in the manner in which practical work is assessed in Ugandan schools, and this is largely due to how practical work is viewed. There is a focus on the development of skills such as observation and handling of materials during experiments. Much of the time, teachers adopt a direct approach to teaching, and do not always conduct practical work as is prescribed in the curriculum (NCDC, 2013). This direct teaching encourages learners to be passive, and the teacher does all the talking. Skills such as categorising of materials using dichotomous keys; mathematical deductions where learners infer their data using descriptive data and graph making as well as their interpretation; and hypothesis formation at O-Level and A-Level, respectively, have been reported to be very vital for developing science understanding (Rubahamya, 2008; Mwesigwa, 2018). However, these are given scant attention.

In summative assessment, the science practical work paper (2 hours) consists of three questions (60 marks). The questions set may be on explanatory notes of new or unfamiliar data to check data interpretation skills or they may use photographs to determine how well learners can interpret information from the photographs, such as cell micrograms. Questions requiring practical laboratory procedures are also set. The total marks for the two sections on practical work and theory is 160. The mark for assessment of practical work is 60 out of a total of 160 marks, and this equates to a sizeable 37.5% of the final mark of the learner in the final (stake) examination.

A case study of practical work at two schools in Uganda

Two government-sponsored schools were investigated for their enactment of practical work in Uganda. Both were mixed-sex secondary schools where they admitted boys and girls at both Ordinary Level (O-Level) and Advanced Level (A-Level). A total of 12 lessons were observed in two schools: six lessons per school. At each school, three lessons were observed for O-Level, and three for A-Level. At both levels, one lesson was observed for each of Physics, Chemistry and Biology. Four research questions were posed: 1) What was the aim of doing practical work in schools? 2) How was practical work conducted in the school? 3) Was the frequency of conducting practical work related to the school curriculum? and 4) How was practical work assessed in the school? Below, I present six lessons from the two schools, with one lesson per subject: Biology, Physics and Chemistry, at each school.

Research context

Two schools in Uganda, Kampala District were purposively selected because they each offered A and O-Levels, and further to this had laboratories for Biology, Physics and Chemistry practical work. In these schools, four teachers, two males and two females, were purposively selected because they taught O and A-Levels. All the teachers were graduates, and two had master's degrees. Their teaching experiences ranged from 10 to 25 years. All the teachers were trained in practical work management and how to design practical activities. Three learners, one from each of Biology, Physics and Chemistry (three from O-Level and three from A-Level), from each school were selected for focus group interviews. All teachers were also interviewed. The interviews with the teachers sought to establish reasons for their actions taken during the practical work, and also to establish the aims of the practical activities. The learners were interviewed on their experiences of the practical work.

Research design and methodology

A case study design (Cohen, Manion & Morrison, 2002) was used because the researcher wanted to gain in-depth understanding regarding the enactment and assessment of practical work in the two secondary schools. Data were collected over a period of one month using lesson observations and interviews. Twelve lessons were observed: six at O-Level and six at A-Level. During lesson observations, the researcher used a time-map event to record what was taking place during the practical lesson. The researcher took photos of groups of learners and the sections of the learners' pages where they filled out the answers. All interviews were audio-recorded: the Personal Interviews (PI) for teachers and the Focus Group interviews (FGI) with the O-Level and A-Level learners. Data from the interviews were transcribed verbatim and analysed thematically (Merriam, 1998; Miles & Huberman, 1994). Using an interpretive process (Schwandt, 2003), the researcher listened to audio-recorded interviews, reviewed the transcripts, and compared and categorised teachers' and learners' statements to generate themes in line with the aim of the study. Four themes were identified: 1) practical work makes a difference in learning science; 2) practical work develops specific skills; 3) practical work enhanced understanding of science concepts (content); and 4) practical work assessment is an integral part of teaching and learning and was in line with the national curriculum. A framework for the analysis of lesson observation and interview data was developed based on the work of Abrahams and Saglam (2010) and Baillie and Hazel (2003) (Table 7.1). The framework covers three aspects: developing explanations and finding solutions; and evaluating data to see their validity and reliability (Osborne, 2011). The aims of practical work suggested by Abrahams and Saglam (2010) are categorised as three broad aims: procedural, conceptual and affective. The specific tenets associated with the aims are: 1) encouraging

Table 7.1 Framework for the analysis lessons observations and interview responses

Level	Description	Tenet (Number)	Category
1	Controlled exercises where the problem, apparatus, procedure and answer are provided.	• Observation and careful recording (1) • Manipulative skills • Preparing learners for assessment (3)	Procedural
2	Structured investigations where problem is given, while the apparatus and method are partly given, and the answer is not given.	• Scientific methods of thought (2) • Preparing learners for assessment (4)	Conceptual
3	Unstructured investigations where only the problem is given, and the apparatus, procedure, and answer are not provided.	• Arousing and maintaining interest (5)	Affective
4	Apparatus, procedure and answer are not provided. It is a project where the problem, apparatus, procedure and answer are not provided.	• Biological, chemical and physical phenomena made real through individual project experiences (6)	Procedural Conceptual Affective

accurate observation and careful recording; 2) promoting simple, common-sense, scientific methods of thought; 3) developing manipulative skills; 4) preparing learners for assessment of practical work; 5) arousing and maintaining interest in the subject; 6) making biological, chemical and physical phenomena more real through actual experiences. Each aim is associated with particular tenets. For example, the procedural aim addresses tenets 1, 3 and 4 in the framework. Observed lessons and interviews were analysed and classified according to levels 1–4 as described by Baillie and Hazel (2003). Level 1 refers to class activities that incorporate controlled exercises where the problem, apparatus, procedures and answer are provided. Level 2 comprises structured investigations where a problem is given, while the apparatus and method are partly given, and the answer is not given. Level 3 consists of unstructured investigations where only the problem is given, and the apparatus, procedure and answer are not provided. Level 4 is a project where the problem, apparatus, procedure and answer are not provided.

Findings

Eight lessons out of twelve were categorised at levels 1 and 2, and four lessons were at level 3 and no lessons at level 4. From the interviews, four themes of doing practical work were identified: 1) making a difference in learning

Table 7.2 List of subjects, classes and topics observed in the two schools in Uganda

Lesson	School	Subject	Level	Topic
1	Mengo Senior Secondary School	Physics	S2 O	Determining the constant P of the meter rule
2	Mengo Senior Secondary School	Biology	S5 A	Food tests and catalyse enzymes in living tissues
3	Mengo Senior Secondary School	Physics	S5 A	Determining the constant k of the resistor marked R
4	Mengo Senior Secondary School	Chemistry	S2 O	Preparation of indicators from flowers
5	Mengo Senior Secondary School	Biology	S2 O	Structure of flowers
6	Makerere College School	Physics	S5 A	Determining the refractive index of a prism
7	Makerere College School	Chemistry	S1 O	Identifying and classifying a substance as soluble, solvent and suspension
8	Mengo Senior Secondary School	Biology	S5 A	Food tests and Vitamin C in foods
9	Makerere College School	Chemistry	S5 A	Back titration
10	Makerere College School	Biology	S5 A	Chemicals of life: Food tests
11	Makerere College School	Biology	S1 O	Identifying different stem modifications and their functions; and describing functions of those modified stems
12	Makerere College School	Physics	S1 O	Determining the density of the provided liquid

science; 2) developing specific skills; 3) enhancing understanding of science content; 4) assessment is an integral part of teaching and learning practical work and was in line with the national curriculum.

All 12 lessons observed are listed in Table 7.2 and six lessons are presented in detail: three from each school covering Biology, Physics and Chemistry. The observed lessons took place in the laboratories of Biology, Physics, Chemistry with workbenches where three to six learners worked together.

Lesson 1 observation

The lesson was in the second year of O-Level Physics at Mengo Secondary School; 60 learners were in the laboratory and one laboratory technician was in attendance. Six learners sat on stools around each table, and each learner had a laboratory manual. The lesson was a double period. It began at 9:10 am and ended at 10 am. The experiment aimed to determine the constant P of the meter rule. For the introduction, the teacher instructed the learners to 1) read the instructions from the practical manual; and 2) set the apparatus: the ruler, wooden wedge block, strings and pins. The learners were asked to balance the ruler on the wooden wedge using different masses ranging from 0.05g to 3.0g and to draw a table of four columns and record on each column the mass (kg), length, l (m), y (m), and 1/y (m-1). The teacher checked if learners were balancing the ruler and recording in the columns. The laboratory technician was on standby to provide materials as needed. In some instances, the teacher identified some learners who could not balance the ruler to get the first data set and they were shown how to balance the rulers. A total of six replica readings were made and recorded in the columns in the laboratory manual.

In terms of the levels of Abrahams and Saglam (2010), the lesson was classified as level 1, that is, a controlled experiment where the problem, apparatus, procedures and answer are provided. The teacher explained what was expected of the learners to experiment. The teacher drew the column format of the table on the chalkboard, and explained and instructed learners to write in each column their findings. The teacher encouraged accurate observation and careful recording. There were comments related to the stake examinations they were to take at the end of the following two years. The teacher prepared learners for answering the examination questions. Thus, the teacher aimed to develop procedural and conceptual knowledge using a structured experiment. Such an approach is most likely to lead to an emphasis on practical skills and following instructions (Hipkins et al., 2002). The teacher contended that the focus was on developing skills in using equipment safely and following instructions.

Lesson 2 observation

The lesson was a double-period 90 minute A-Level Biology class for senior five at Mengo Secondary School. It started at 2:20 pm and ended at 3:40 pm. There were 55 learners in the class, and there was one laboratory technician. The teacher instructed the learners to 1) read the instructions from the practical manual; 2) collect the chemicals needed (hydrochloric acid, DCPIP, iodine solution and hydrogen peroxide); and 3) use the knife, Irish potatoes, mortar and pestle as per instruction from the laboratory manual. The teacher instructed learners to fill in the columns from the table in the laboratory manual: column 1, test, column 2, observation, and column 3, conclusion.

Learners in the first part of the practical test had to test the type of food present in a potato using iodine solution, Benedict's solution (hydrochloric acid, sodium hydroxide and copper sulphate solutions). Learners in the second part of the practical work used Dichlorophenolindophenol (DCPIP) solution. DCPIP is blue with maximal absorption at 600 nm; when reduced, DCPIP is colourless. Learners peeled the potatoes and cut four cubes of 3cm × 3cm × 3cm and ground them in the mortar to obtain extracts which were filtered and collected in the test tubes. Also, they cut a filter paper: 0.5cm by 0.5cm. The filter paper was dipped into hydrochloric acid, hydrogen peroxide and introduced into the potato extract in four test tubes labelled A1, A2, A3 and A4. The learners recorded the time for the paper to sink and to rise again.

Learners followed the instructions by filling in the columns in the laboratory manual with the procedures in the test column; their observation in the observation column; and their deductions in the conclusion column. At various intervals, the teacher reminded the learners to follow instructions. The teacher also reminded them to think about why each step of the experiment was important. This information was necessary because the learners had to state the aim of the experiment and suggest another way of doing the same experiment. In the end, learners were to state the aim of the experiment. The conclusions were to be in line with what they observed. The teacher moved around and marked a few books and assisted those in need, along with two laboratory assistants. The teacher also looked for concepts learners exhibited during practical work and made announcements from time to time to guide the process.

At the end of the experiment, the teacher and learners reflected on the process and discussed new ways of doing the same experiment. The teacher asked a few learners the aim of the experiments. Other learners challenged the answers given. Learners debated the accepted aim of the experiment. They concluded that the aim was to test the effect of the active substance in the specimen. The teacher asked the learners to explain what would be tested if the number of drops of acid were increased. Furthermore, the following questions were posed: How would the increase in acid affect the rate of the reaction? What would happen if the temperature was varied between $25°C$ and $40°C$? Various answers were given by different learners, and after the discussion, the teacher justified the acceptable answer. Finally, the teacher elaborated upon what it means to explain in a practical lesson. He emphasised three things to include when asked to explain something in practical work. These are: 1) write down what happened; for example, the paper sank and rose to the surface; 2) write down how it happened by stating what process led to what was observed, such as the active substance hydrogen peroxide decomposed at the bottom of the liquid in the test tube; and 3) write down the reasons for what happened, for example the sinking of the filter paper to the bottom of the test tube.

In this lesson, the teacher and the learners operated at levels 2 and 3, where the aim and answers and procedures are not provided. The researcher noted that the teachers in this lesson aimed to develop not only procedural and conceptual knowledge, but also to address the affective domain. Learners debated the aim of the experiment because the teacher did not provide the aim from the beginning. The teacher's actions were intended to encourage accurate, critical thinking, and he challenged the learners with other possible scenarios that could form part of other experiments. The affective domain is supported because many learners were amazed to observe the sinking and the rising as they timed the duration.

Lesson 3 observation

The lesson was a double-period 90 minute A-Level Chemistry class for senior five at Makerere College School. It started at 2:10pm and ended at 4:30 pm. There were 50 learners in the class and two laboratory technicians. The learners were in groups of three to four per table and were all seated on stools. The aim of the experiment was to determine the Relative Atomic Mass (RAM) of a metal M in its hydroxide MOH. The procedures were 1) measure 0.6 g and dissolve in $250cm^3$ and make up to the resultant solution; 2) add 2 drops of phenolphthalein indicator and titrate the solution with FA_3 solution until there is a change in colour showing the endpoint. The teacher cautioned learners to be watchful to notice the endpoint because there is a danger of overrunning the titre. A few learners asked for clarifications here and endeavoured to make sure they did not make unnecessary errors. Three replicas were titrated, and the average was taken as the value. The RAM of the metal M was calculated from the titre.

The lesson focused on procedures and conceptual understanding, and was categorised at levels 2 and 3 (Baillie & Hazel, 2003). The teacher emphasised the importance of accurate measurements of solids, liquids and during titration to come up with the precise RAM of the metal M. Furthermore, the teacher emphasised that each member of the group had a role to play and that any member failing to fulfil the role assigned to him by the group would jeopardise the mark of that group. Consequently, the class got the RAM of M as expected.

Lesson 4 observation

The lesson was a double-period 90 minute A-Level Physics class for senior five at Makerere College School. It started at 2:10 pm and ended at 4:30 pm. There were 50 learners in the class and two laboratory technicians. The experiment aimed to determine the refractive index of a prism. All learners had a board and a 60^0 prism. During the introduction, the teacher instructed the learners to 1) read the instructions from the practical manual; and 2) set the apparatus: the ruler, paper, drawing pins, protractor, cello tape and pencil. A set of detailed procedures were outlined to the learners. The learners performed the experiment, calculated the refractive index, drew graphs and drew the figures. The teacher marked the learners' calculations from the practical work. The teacher emphasised accuracy in drawing the graphs, and the importance of writing the title of the graph and labelling all the axes. The specific tenets addressed were: 1) encouraging accurate observation and careful recording; 2) promoting simple, common-sense, scientific methods of thought; 3) developing manipulative skills. Thus, the lesson was at levels 1 and 2 in terms of the framework by Baillie and Hazel (2003).

Lesson 5 observation

The lesson was a double-period O-Level Biology class for senior two at Mengo Secondary School. There were 60 learners and one laboratory technician. Six learners sat on stools around each table, and each learner had a laboratory manual. The double period began at 2:10 pm and ended at 3:20 pm. The experiment was to outline and describe the differences between specimen Q and R provided. For the introduction, the teacher instructed the learners to 1) read the instructions from the practical manual; 2) observe the specimen Q and R; 3) state agents of pollination; 4) suggest reasons for the pollination agents for Q and E in 3) above; and 5) remove the petals and sepals from specimens Q and R, and label the rest of the parts. During the lesson, the teacher emphasised careful observation of the specimens and careful drawing of the parts. For each specimen, the teacher emphasised describing the role of agents of pollination. He also reminded learners that understanding the procedures would assist them to perform well in the final examination.

Learners' actions observed were matched to tenets (Abrahams & Saglam, 2010). These actions included accurate observation and careful recording; learners followed instructions to learn procedures and to develop manipulative skills; learners were prepared for the final practical examination. Thus, considering the learners and the teacher's actions in terms of the framework by Baillie and Hazel (2003), the lesson was at levels 1 and 2.

Lesson 6 observation

The lesson was a double-period 2 hour A-Level Chemistry class for senior five at Makerere College School. There were 60 learners in the laboratory and one laboratory technician. Six learners sat on stools around each table, and each learner had a laboratory manual. The lesson began at 11:00 am and finished at 1:00 pm. The experiment was to determine the percentage purity of a metal carbonate using the standard solutions of hydrochloric acid and sodium hydroxide. For the introduction, the teacher instructed the learners to 1) read the instructions from the practical manual; and 2) to take samples of the chemicals: solid Z, which is an ore of a metal carbonate MCO_2 with a molecular mass of 100g; FA1, which is 1M hydrochloric acid solution; and FA2, which is 1M sodium hydroxide solution. The following procedures were outlined: a) Measure 60cm^3 of FA1 into a clean beaker followed by 140cm^3 of distilled water. Label the resultant solution FA3; b) Transfer exactly 50cm^3 of FA2 into a clean 250cm^3 volumetric flask and add distilled water up to the mark. Label the resultant solution FA4; c) weigh 1.5g of Z into a clean beaker and accurately add 80cm^3 of FA3. Keep the reaction mixture until no further observable reaction. Add 120cm^3 of distilled water and label the resultant FA5, d) Pipette 20cm^3 of FA5 into a conical flask followed by 2 to 3 drops of phenolphthalein indicator and titrate the resultant mixture with FA4 till the endpoint. Repeat the titration three times to obtain consistent results and record the results in a table. The average volume used was calculated by taking the final volume and subtracting it from the initial volume and dividing by 3. Finally, learners calculated the percentage of the amount of impunity of the metal ore and determined the percentage purity. The teacher at various stages cautioned learners to be accurate at all stages of the work. He cautioned learners not to run too much liquid from the burette and miss the endpoint, which would result in errors in the impurity present in the ore. He reminded them to note the steps in their calculations because such steps are likely to be tested in the stake examinations.

In this lesson, using tenets from the framework of Baillie and Hazel (2003), learners followed instructions to master the procedures and to use their science concepts to work out the solutions. According to Abrahams and Saglam (2010), the lesson covered procedural and conceptual categories. Furthermore, learners were prepared for the eminent stake examination in the following year. Thus, the lesson was at levels 2 and 3.

Interviews with teachers and learners

In order to understand the teachers' purpose of conducting practical work, the researcher used Personal Interviews (PI). To obtain the learners' views regarding how practical work was conducted in the schools, the researcher used Focus Group Interviews (FGI) with six learners per group. Teachers are coded as T1, T2, T3 and T4. Learners are coded as L, individual learners identified by a letter: A, B, C… K and the focus group identified with 1, 2, 3 and 4. For example, LA1 means the learner is A and belongs to focus group 1, while LB2 means the learner is B and belongs to focus group 2.

Teachers were adamant that practical work makes a difference to the learners' education. They clearly stated that the learners who miss practical work are distinctly different from those who have done practical work. Similarly, the learners had a view that if they do not do practical work, they would miss out on fun and understanding content (theory). The following excerpts from teachers and learners, respectively, represent these perspectives.

T2: The person who is taught with practical work is far better than the person who is not taught with practical work. For example, in practical work learners improve in their reasoning ability and learn how to interpret things and of course that learner will be different from the other one who has been taught theoretically.

LA2: We would miss a lot, because if I look at the news, there are those learners who meet apparatus for the first time in the final examination room, but here we are given practice every week. We are used to the apparatus, so it is easy, going to be easy in the examinations.

LB2: And even if different apparatus are used and someone says something, I can get the idea. If you bring a question including a meter bridge, I have

seen a meter bridge before so when you put down the terms potenti-
ometer and meter bridge I know how to differentiate them, so I know
how to use different apparatus and also record readings from them.

Teachers contended that the time they are provided for practical work meets
the requirement of the curriculum. Each science subject meets once a week for a
double period throughout the term. These are presented in the following excerpts:

T4: We are in line with the curriculum, because we teach the theory along
with the practical, in mixed school. Very few have practicals in senior one
like the way we do it here at Makerere.
T2: In relation to the NCDC (National Curriculum Development Centre
[NCDC]), they provided the syllabus and after providing the syllabus, we
follow it up and we do experiments according to the NCDC.

All the teachers stated that assessment of practical work features strongly, and
they regard this as good preparation for the final practical work examination at
the end of the term and at the end of the year as well as at the end of four years
for O-Level and two years for A-Level. Learners also recognise the importance
of these practical lessons for the final high stakes examination. This is apparent
in the following excerpts:

T4: We give two sessions, in the end of term examinations we give them a
practical paper and theory paper independently, in classroom... assessment
after the practical they have questions in the practical book so they read the
questions in the books and answer them. The learners benefit in two ways:
theory paper and practical paper. At the end of the term there is a practical
paper exam, which is the unique venture of Makerere College School.
LD3: These practicals, they are examinable at the Uganda National Examination
Board (UNEB), so when you do not do them right now, when the time
comes to do them at UNEB, you might panic not knowing what to do
and you just know the cram work of the book.

The teachers regard formative and summative assessment practical work as
bring related, with formative assessment preparing learners for summative
examinations. This is illustrated in the following two teachers' excerpts:

T2: Arrhh... The difference maybe there in relation to the Uganda National
Examination Board (UNEB), because the final assessment is set by UNEB,
but if it is term work normally what they keep on doing in the class is
totalled with the end of term exam. We give say 25% practical work class
work and 75% comes from the final theory examination.
T3: It is similar to the final except that the final assessment like when you
maybe around... we are going to introduce another animal that is a toad

or a rat, in the final they are expected to dissect and draw, but the dissecting procedure is not marked, and to me I feel it is not good. The assessment is not complete. It is like cooking food then it is not tested there and then. Then somebody marks the theory of cooking that food.

Skills development is flagged as being important in any practical activity. This is highlighted by both teachers and learners, and is reflected in the following excerpts:

T1: Practical work we emphasize observation skill, recording, recording results, results.

T4: The best we do, we want stimulation, developing skills, skills like observation, recording, etcetera. So, we give a foundation, we give a basis for an examination.

LA1: Okay, okay. Skills of plotting graphs even following instructions from the teachers.

LB2: You can have different apparatus and in what way they can be used in case I have…

Teachers believe that practical work is important to enhance understanding of science content. For that matter, teachers were convinced that doing practical work was necessary to reinforce the content learners study during the theory lessons. Learners also see the value of doing practical work to improve their understanding of science content. This is apparent in the following excerpts:

T1: Practical work is to improve the learners' content (theory).

T4: We also prepare them for their future career.

LA1: Helps us to examine practical theory such that we do not just get things theoretically, so that we know where they come.

LA2: Improves on our understanding of theory because you are doing what you have been studying, so you get to know more.

Teachers stated that they did not set tasks where learners design their own experiments due to lack of time and learners confirmed the teachers' statements. These views are represented in the following excerpts:

T4: Aaah… Unfortunately, no. The level of senior 1 and senior 2, we do not have that design.

T1: In senior 3 and senior 4, one could ask them, but there is always not enough time for that. After all, learners will not be asked to design in senior 4 final examinations.

Discussion of findings

Practical work is an integral part of teaching and learning science in Ugandan secondary schools. It makes a difference in the mastering of content and in developing specific skills. Both teachers and learners contend that practicals enhance content, skills, create interest in learning science and prepare learners to pass their final (stake) examinations. Out of 12 lessons, eight of them were in levels 1 and 2. The other four were at levels 2 and 3, and none was in level 4. This observation suggests that most of the teachers operated at level 1, where they provided the problem, apparatus, procedure and learners worked out the answer, and level 2 by providing the problem, apparatus and part of the method and learners worked out the solutions (Baillie & Hazel, 2003).

All three science subjects, Biology, Physics and Chemistry, have one practical work session per week. A learner studying all three science subjects must attend laboratory work three times a week. In ten weeks of study, the learner attends 30 sessions of practical work per term. Thus, for the three school terms in the Ugandan curriculum, the learner attends 90 sessions of practical work in one year. Considering six years of secondary education in Uganda, learners attend 540 laboratory work sessions before enrolling at university. Thus, laboratory work is taken seriously in secondary schools in Uganda. The government employs both the teachers and science laboratory technicians. The role of the technician, in the laboratory, was to attend to faulty apparatus. One school had one laboratory technician per subject while the other school had two laboratory technicians per subject: Biology, Physics and Chemistry. Technicians prepare the materials for experiments under the guidance of the subject teacher. They assist learners together with the teacher during practical work sessions.

During practical work sessions, teachers and laboratory technicians scaffold learners. They do this by monitoring progress, checking to see that instructions are being followed, and when learners run into difficulty they are offered hints to overcome the challenges. For the O-Level phase, learners are expected to have mastered practical skills to be used at the end of the O-Level phase stake practical examinations. The stake practical work examination is set by the external body, the Uganda National Examination Board (UNEB). It means that all learners in Uganda are assessed on how well they have mastered practical skills during the four years of study.

Similarly, at the end of two years of A-Level study, learners take a stake practical examination set by UNEB to assess how well learners have mastered practical skills during the study. The final mark for the learners consists of 25% from practical work. The other 75% is achieved from the theory work. In the practical exams, learners are assessed for skills such as following instructions, setting up apparatus and performing experiments.

Furthermore, the skills gained in practicals are not well spelt out to the learners. This became evident when some learners in the interviews admitted that they did not know how skills such as observation, making inferences, drawing, and working in a team could be regarded as being relevant. When learners fail to recognise the objectives of experiments, there are negative implications for learning (Schauble et al., 1995). Some learners considered laboratory work as manipulating apparatus and not a process of developing concepts and ideas (White, 1998). While most learners admitted that practical skills are necessary to acquire, they lacked an understanding of the role of knowledge and skills acquired through practical work beyond the school. The value of practical work for societal application needs to be more explicitly stated by teachers. For example, knowledge on the food types tested in the class can be used to determine the foods to buy in order to have a balanced meal at home. Also, the skill of measuring liquids can be used when measuring pesticides to spray on crops, flowers or on cattle to free them from ticks.

At A-Level, unlike at O-Level practical work, learners were metacognitively engaged by being prompted by the teacher to reflect on their actions when they did the experiment. Thus, learning was enhanced by identifying, monitoring and regulating their thinking and learning process (Gobert & Clement, 1999). An example of this was when A-Level learners who used a galvanometer were asked at the end of the experiment to explain how they would design the same experiment using their prior knowledge but using different apparatus. Some learners suggested a meter bridge as an alternative to what their teacher had provided in the laboratory manual. Thus, metacognitive strategies included connecting prior knowledge to new knowledge, selecting thinking strategies and monitoring the progress during problem-solving (Gobert & Clement, 1999).

Conclusion

In teaching practical work, teachers need to create learning experiences where learners can deepen their understanding and learn to think critically about the world around them. This chapter has presented empirical evidence on how practical work is conducted in selected schools in Kampala District in Uganda. The importance given to practical work in science learning is evident from the fact that lessons on practical work are timetabled. The school timetable showed that practical work takes place once a week for each of the three subjects. This translates to 25–30% of the total school periods, and suggests that Ugandan secondary schools take science practical work very seriously. It is no wonder, as in the final (stake) examination, learners are expected to write a practical paper for each of the three science subjects. The availability of laboratory manuals in schools shows that teachers, learners and the school administration promote practical work in schools. Finally, interviews conducted with learners show that learners regard laboratory work as fun. The lack of assessment and awarding marks for performing practical work, the lack of student-designed tasks and the lack of ICT-integrated practical lessons are the missing links of laboratory work in Uganda, and these need urgent attention.

Aina (2013) contends that teachers should use local materials from the environment to teach Physics. In Uganda, the practicals of Biology and Chemistry used a few available local materials from the community. Materials included plastic cups, pins, rubber bands, strings, potatoes, flowers, tendrils, onions, and other plant stems. By doing experiments at the observed school using various local materials, learners were engaged in understanding the world around them and in mastering the subject content as well as the process. In 2005, shortly after the Ministry of Education and Sports (MoES) announced a programme of Universal Secondary Education, all students at O-Level were expected to study science. This decision resulted in many learners doing science, resulting in a shortage of instructional resources (Liang, 2002; MoES, 2015). As many as 120 learners were in one of the O-Level science laboratories. The use of low-cost resources may be a solution to this challenge.

Information Communication Technology (ICT) has permeated all sectors of life, including education, and it can be one of the approaches to minimise the shortage of resources in schools (Kibirige, & Tsamago, 2019). In one of the schools, there were plans for learners to view a video in O-Level Biology during the practical lesson; unfortunately, there was no electricity. Nevertheless, it is critical to venture into studies regarding the use of ICT in practical work. It is imperative to study the teachers' views regarding the use of ICT, such as computer simulations, and also to identify the teachers' competence in using technology to teach practical work and science topics.

References

Abrahams, I., & Saglam, M. (2010). A study of teachers' views on practical work in secondary schools in England and Wales. *International Journal of Science Education*, 32 (6), 753–768.

Aina, K. J. (2013). Instructional materials and improvisation in physics class: Implications for teaching and learning. *Journal of Research and Method in Education*, 2(5), 38–42.

Altinyelken, H. K. (2010). Pedagogical renewal in sub-Saharan Africa: The case of Uganda. *Comparative Education*, 46(2), 151–171.

Baillie, C., & Hazel, E. (2003). *Teaching materials laboratory classes*. Liverpool: The UK Centre for Materials Education.

Bell, P., & Linn, M. C. (2000). Scientific arguments as learning artifacts: Designing for learning from the web with KIE. *International Journal of Science Education*, 22(8), 797–817.

Cohen, L., Manion, L., & Morrison, K. (2002). *Research methods in education* (5th ed.). London: Routledge Falmer.

Gobert, J., & Clement, J. (1999). The effects of student-generated diagrams versus student-generated summaries on conceptual understanding of spatial, causal, and dynamic knowledge in plate tectonics. *Journal of Research in Science Teaching*, 36(1), 39–53.

Hipkins, R., Bolstad, R., Baker, R., Jones, A., Barker, M., Bell, B., & Haigh, M. (2002). *Curriculum, learning and effective pedagogy: A literature review in science education*. New Zealand: Ministry of Education.

Kibirige, I., & Tsamago, H. (2019). Exploring Grade 10 learners' conceptual development using computer simulations. *Eurasian Journal of Mathematics, Science & Technology Education*, 15(7), 1–17.

Liang, X. (2002). *Uganda post-primary education sector report*. Washington, DC: World Bank.

Merriam, S. B. (1998). *Qualitative research and case study applications in education*. San Francisco: Jossey-Bass Publishers.

Miles, M. B., & Huberman, A. M. (1994). *Qualitative data analysis* (2nd ed.). Thousand Oaks, CA: SAGE.

Ministry of Education and Sports (MoES) (2006). *The Ugandan primary teacher education curricula review*. Kampala: The Uganda Ministry of Education and Sports.

Ministry of Education and Sports (MoES) (2008). *The Education and Sports Sector Annual Performance Report (ESSAPR) covering financial year 2007/08*. Kampala: The Uganda Ministry of Education and Sports.

Ministry of Education and Sports (MoES) (2015) *Education, science, technology and sports sector annual performance report, 2014/15*. Kampala: The Uganda Ministry of Education and Sports.

Mwesigwa, D. A. (2018). *A simplified approach to O-Level chemistry practicals*. Kampala: Jescho Publishing House. National Academic Press.

National Curriculum Development Centre (NCDC) (2006). *The National Primary School Curriculum for Uganda*. Teacher's guide, Primary 1. Kampala: The National Curriculum Development Centre (NCDC).

National Curriculum Development Centre (NCDC) (2013). *Secondary Curriculum*. Retrieved 13 March 2019 from www.ncdc.go.ug/content/o-level-curriculum.

O'dama, M. (2013). Triumph and prosperity of education in Uganda. In T. S. Mwamwenda & P. Lukhele-Olorojunj (Eds), *Triumph and prosperity of education in Africa* (pp. 539–565). Pretoria: Africa Institute of South Africa.

Osborne, J. (2011). Science teaching methods: A rationale for practices. *School Science Review*, 98(343), 93–103.

Rubahamya, J. B. (2008). *UACE Biology practical guide for P530/3*. Kampala: Grapet Production.

Schauble, L., Glaser, R., Duschl, R. A., Schulze, S., & John, J. (1995). Students' understanding of the objectives and procedures of experimentation in the science classroom. *Journal of the Learning Sciences*, 4(2), 131–166.

Schwandt, T. (2003). Three epistemological stances for qualitative inquiry: Interpretive, hermeneutics, and social constructivism. In N. Denzin & Y. Lincoln (Eds.), *The landscape of qualitative research: Theories and issues* (pp. 292–327). Thousand Oaks, CA: Sage.

Ssempala, F. (2017). Science teachers' understanding and practice of inquiry-based instruction in Uganda. Unpublished doctoral thesis, Syracuse University, United States of America.

Stroupe, D. (2015). Describing 'science practice' in learning settings. *Science Education*, 99(6), 1033–1040.

Torrance, H. (2012). Normative assessment at the crossroads: Confirmative, deformative and transformative assessment. *Oxford Review of Education*, 38, 323–342.

van der Graaf, J., van de Sande, E., Gijsel, M., & Segers, E. (2019). A combined approach to strengthen children's scientific thinking: Direct instruction on scientific reasoning and training of teacher's verbal support. *International Journal of Science Education*, 41(6), 713–738.

Chapter 8

Science and engineering practices coverage in science practical work

Analysis of Zambia's Integrated Science Curriculum materials

Vivien Mweene Chabalengula and Frackson Mumba

Introduction

At the beginning of the 21st century, science curricula reforms of various countries affirmed that practical work in science education is central to the development of scientific literacy (Abrahams, 2017; Akuma & Callaghan, 2019; Allen, 2012; Ferreira & Morais, 2014; Lunetta & Hofstein, 2007). Examples of such science reforms emphasizing practical work include the National Science Education Standards in the USA (NRC, 1996), the Nuffield Curriculum in the United Kingdom, the Biological Sciences Curriculum Study in the USA (Lunetta et al, 2007), and the Australian National Curriculum (Kidman, 2012). Similarly, Zambia's current Integrated Science Curriculum (ISC) at junior high school level reaffirms that practical work is invaluable for teaching and learning science concepts, and for use as contexts for applying knowledge across curricula and in everyday-life experiences (MESVTEE, 2015). However, each nation tends to hold different conceptions of what constitutes practical work. For example, in the USA, the National Science Education Standards refer to lab-based/inquiry learning activities as contexts through which students develop knowledge and understanding of scientific ideas, as well as an understanding of how scientists study the natural world. In Zambia, the Integrated Science Curriculum defines practical work as an overarching term that includes hands-on experiments, teacher demonstrations, mind-on discussions and activities, oral activities, and field/project work (MSEVTEE, 2015). Though there is no agreed-upon definition of what constitutes practical work, some authors (e.g. Abrahams & Millar, 2017) assert that it generally refers to learning experiences in which students interact with and manipulate real materials/objects/equipment or secondary sources of data to observe and understand the natural world. In this chapter, we will refer to practical work as it is conceived in Zambia because we will analyse Zambian curriculum materials.

Science education literature highlights several fundamental benefits of science practical work, which include: enabling students to learn with understanding of science concepts and engaging them in the process of constructing knowledge by doing science (Abrahams & Millar, 2008; MSEVTEE, 2015); providing meaningful learning experiences/contexts for students to explore science phenomena, perform experiments, and acquire scientific and critical thinking skills (Ferreira & Morais, 2014); helping students make links between the real world of objects, materials and events, and the abstract world of thought and ideas (MESVTEE, 2015); helping students appreciate that science is based on evidence and to acquire science inquiry skills (Abrahams & Millar, 2008); and that students find practical work motivating and enjoyable as compared to other science teaching and learning activities (Abrahams & Millar, 2008; Allen, 2012).

Despite the highlighted benefits of science practical work, science education research shows mixed findings about its effectiveness in facilitating students' understanding of scientific phenomena and development of inquiry skills (e.g. Abrahams, 2017; Chabalengula & Mumba (2012; Gott & Duggan, 2007). For example, while Hewson and Hewson (1983) reported a significant enhancement of students' conceptual understanding among students who had received practical-based instruction compared to those who received a traditional non-practical instruction, Mulopo and Fowler (1987) reported no significant differences. Hodson (1991) found that many students stated that practical work contributes little to their learning of science, and concluded that practical work is ill-conceived, confused, and unproductive as practiced in many schools. Lazarowitz and Tamir (1994) found that practical work offers no significant advantage in the development of students' scientific conceptual understanding. More evidence from Abrahams and Reiss's (2012) evaluation of the national project designed to improve the effectiveness of practical work in England revealed that practical tasks were less effective in enabling students to use the intended scientific ideas to understand their actions and reflect upon the data they collected, and that teachers consistently failed to incorporate activities that would make links between the observations and the scientific ideas that they were designed to illuminate. Other authors have concluded that practical work activities tend to be mostly teacher-driven or cookbook in nature where learners simply follow the provided experimental procedures with little thought and purpose (e.g. Akuma & Callaghan, 2019; Kidman, 2012). A recent study by Abrahams (2017) found that many students were unable to articulate what they learned from the practical task, or the reason they undertook the task. In the Zambian context,

Our previous study that surveyed Zambian teachers' conceptions of inquiry and examined the inquiry levels in the national high school science curriculum materials (syllabi, textbooks and practical exams) revealed that teachers held narrow conceptions of inquiry as they perceived giving more priority to evidence gathering and explaining the evidence, with less emphasis on justifying the explanations and connecting the explanation to scientific knowledge, while curriculum materials put much emphasis on lower inquiry tasks and skills which are confirmation and structured inquiry (Chabalengula & Mumba, 2012). Similarly our own personal experiences as science educators in Zambia inform us that practical work is conducted in a "vacuum", without a sense-making context in which students can apply their knowledge and evaluate the knowledge from investigations. Typically, practical work in Zambia tends to engage students in investigating relationships between variables or test hypotheses such as how the mass of an object affects its volume, but students are not involved in relating these empirical explorations to real-world problems. (Examples of what practical work activities look like in Zambia are provided in the "Methodology" section.) As such, students and teachers alike do not see any meaningful end goal for doing practical work. Our experiences are supported by Osborne and Quinn (2017), who found that many inquiry-based science classrooms emphasize investigative procedures without any links to real-world problems.

In light of the mixed findings about the benefits of science practical work, Clackson and Wright (1992, p. 40) succinctly summarized the situation as follows: "Although practical work is commonly considered to be invaluable in science teaching, research shows that it is not necessarily so valuable in science learning". To remedy the situation, some researchers have proposed some solutions;., such as: introducing students to the relevant scientific concepts prior to them undertaking any practical work if the task is to be effective as a means of enhancing the development of their conceptual understanding (Hodson, 1991); positioning practical work as a *link/bridge* between previously taught scientific concepts and subsequent observations (Millar et al, 1999); and engaging students in practical work experiences in ways that optimize the potential of practical work activities as a *unique and crucial medium*, which promotes the learning of science concepts, procedures, processes, and other important goals in science education such as linking knowledge to everyday-life situations (Lunetta et al, 2007). The idea of a *link/bridge/unique medium* for practical work has recently been incorporated by other science education systems, particularly the USA where pedagogical innovations in practical work are moving beyond simply designing investigations and testing hypotheses to the fuller articulation of inquiry in the form of science and engineering practices outlined in the K-12 Science Education Framework (i.e. asking questions for science and defining problems for engineering; developing and using models; planning and carrying out investigations; analyzing and interpreting data; using mathematics and computational thinking; constructing explanations for science and designing solutions for engineering; engaging in argument from evidence; and obtaining, evaluating, and communicating information). These are designed to enable students to investigate and make sense of scientific phenomena by building and applying explanatory models and by designing solutions to world problems (NRC, 2012). Similarly, Zambia's Integrated Science Curriculum (MESVTEE, 2015) emphasizes positioning practical work as a *learning context* that can reinforce linking science knowledge, inquiry skills, and knowledge application to life experiences, cross-cutting issues and emerging challenges that cut across the curriculum to ensure the holistic development of a learner, and to ensure learners gain competencies in mathematical, scientific, technological, communication, and general life skills such as critical thinking and creative reasoning. However, there is a dearth of research evidence in Zambia about the extent to which practical work as a learning context is facilitating the aforementioned emphases, particularly the triad among concept understanding, procedural skills, and ability to evaluate the investigation procedures and the scientific claims being generated.

Therefore, the purpose of this chapter is twofold: (a) to analyze the extent to which practical work activities address the science and engineering practices, using the K-12 Science Education Framework as an analysis guide. (b) To propose a design matrix for science practical work activities based on the findings. We utilized the K-12 Science Education Framework because it outlines the science and engineering practices that are a kind of improved scientific inquiry process that articulates more clearly what successful inquiry looks like when it results in conceptual and procedural understandings (Schwarz et al, 2017). From a Zambian perspective, Chabalengula and Mumba (2012) found that many science teachers believed that if students were not involved in some kind of hands-on activity or experiment, then it was not inquiry. Therefore, using the science and engineering practices framework will help educators understand that students can engage in inquiry in various modalities such as minds-on activities. Specifically, we analyzed Zambia's new Integrated Science Curriculum national syllabi and textbooks that were introduced in junior high schools in 2015. Our rationale for analyzing the syllabi and textbooks is that they are the main sources of science teaching and learning for both teachers and students, and are also used as conceptual structure guides when examiners are preparing national science exams. As such, the findings will be of value to a wider Zambian science education community. The research questions that guided this study were: (a) To what extent are science and engineering practices covered in Zambia's Integrated Science Curriculum syllabi? (b) To what extent are science and engineering practices covered in Zambia's Integrated Science Curriculum textbooks? (c) What is the coverage of science and engineering practices across practical activities in discipline-specific concepts (i.e. life science and physical sciences)?

This study is significant because as current trends and pedagogical innovations in practical work are moving beyond simply designing investigations and testing hypotheses to more complex understandings in the form of science and engineering practices, there is a need to know the extent to which practical work activities integrate these practices. This knowledge is critical, as it would inform science educators, curriculum developers, and teacher education programs on which practices to consider during curriculum design and development of learning experiences for students. Furthermore, an understanding of the current status of practical work would be helpful in furthering the development of meaningful practical work learning experiences among all students in Zambia and other countries.

Zambia's Integrated Science Curriculum and science practical work activities

In Zambia, junior high school science education has a national curriculum, and lasts for two years, from Grade 8 to Grade 9. Previously, the compulsory science course in junior high school was Environmental Science but was recently revised, updated, and replaced by the Integrated Science course in 2015 (MESVTEE, 2015). The Integrated Science Curriculum covers a range of concepts from biology, chemistry, and physics disciplines. According to the Zambian Education Curriculum Framework (MESVTEE, 2015), the junior high school Integrated Science Curriculum was revised and updated in order to (a) follow the outcomes-based education (OBE) learning approach, which links education to life experiences as it gives learners skills to access, criticize, analyze and practically apply knowledge; (b) to integrate cross-cutting issues and emerging challenges that cut across the curriculum to ensure the holistic development of a learner; and (c) to ensure learners gain competencies in mathematical, scientific, technological, communication, and general life skills such as critical thinking, creative thinking, reasoning, manipulating, as well as attitudes and values. These recent changes to the junior high school Integrated Science Curriculum materials perfectly render themselves to be evaluated for coverage of current trends and pedagogical orientations, particularly the integration of science and engineering practices.

The Integrated Science Curriculum has national syllabi and two sets of textbooks (a teachers' guide and a learners' book), which were written by Zambian science educators and published and approved by the MESVTEE in 2015. (The full references for the syllabi and textbooks are provided in the reference list). These textbooks and syllabi are the main sources of science teaching and learning for both teachers and students in junior high schools. These are also used as guides when examiners are preparing national exams. Each science teacher is given the textbook and a copy of the syllabus as a guide for scope and depth of the content to be taught. All students are given copies of the Integrated Science textbooks for their current grade, and are required to return the books at the end of each grade. The minimum learner–teacher contact time to cover the Integrated Science Curriculum is six 40-minute periods per week. There are two single lessons and two double lessons per week.

The syllabi for Grades 8 and 9 are divided into topics, and each topic is introduced with its learning outcomes that highlight the content knowledge, inquiry skills, and values that a learner is expected to master and demonstrate as a result of the learning experience. The inquiry skills statements are the units of analysis in this chapter because they are directly related to practical work. The Grade 8 syllabus has seven topics, which correspond to the seven chapters in the learner's textbook. The topics outlined in the Grade 8 syllabus are: The Human Body, Health, The Environment, Plants and Animals, Materials, Energy, and Composition of Air. The Grade 9 syllabus has ten topics, which also correspond to the ten chapters in the learner's textbook. The topics outlined in the Grade 9 syllabus are: The Human Body, Health, The Environment, Conservation of Animals and Plants, Plants, Chemical Reactions, Light, Electric Circuits and Pressure, Energy and its Conservation, and Communication.

The Grades 8 and 9 textbooks are titled *Integrated Science 8* and *Integrated Science 9*, both published by Pearson. The Grade 8 and Grade 9 textbooks have seven and ten chapters, respectively, which correspond to the seven and ten topics in their respective syllabi (see list of topics above). In both textbooks, each chapter has various activities in the form of hands-on activities (such as experiments), and minds-on activities (e.g. discussion questions). The hands-on activities are the units of analysis in this chapter because students are required to practically conduct the experiments. Typically, practical work activities begin with the aim, materials/equipment required, the step-by-step procedure, and questions for students to answer that may require them to either explain, analyze, or interpret data. In some cases, the data analyses, interpretations, and explanations are provided for students at the end of each experiment. For actual examples of what practical activities look like, refer to Table 8.3 in the "Data Analysis" section.

Methodology

Data sources and units of analysis

The data sources were two syllabi (each for Grades 8 and 9), and two learners' textbooks (one for Grades 8 and 9). As already stated in the previous section, the units of analysis were the inquiry skills statements and hands-on activities in the syllabi and textbooks, respectively. Zambia's Junior High School Integrated Science Curriculum covers both biology and physical science discipline-specific hands-on activities. In this chapter, we define discipline-specific activities as those that either have a biology or physical science emphasis. For example, the hands-on activities under the topic "Human Body" are categorized as biology activities, while those under the topic "Chemical Reactions" are categorized as physical science activities. Table 8.1 shows the profiles of the syllabi and textbooks with respect to the science topics covered, number of inquiry statements in each syllabus, number of hands-on activities in each textbook, as well as whether the topic has a biology or physical science emphasis.

Content analysis framework

The K-12 Science Education Framework developed by the National Research Council (2012) was used as the analysis framework. The framework outlines eight science and engineering practices, each with an **anchoring** *phrase(s)* that served as analysis guides during coding. Table 8.2 shows concise descriptions of the

Table 8.1 Profile of syllabi and textbooks analyzed

Data source	Science topics covered	Number of units of analysis per topic in each source		Total units of analysis
Grade 8 syllabus & textbook		Syllabus: Inquiry statements	Textbook: Hands-on activities	
	The Human Body (B)	5	3	8
	Health (B)	2	1	3
	The Environment (PS)	3	3	6
	Plants and Animals (B)	10	4	14
	Materials (PS)	22	12	34
	Energy (PS)	11	11	22
	Composition of Air (PS)	3	8	11
	Total	56	42	98
Grade 9 syllabus & textbook		Syllabus: Inquiry statements	Textbook: Hands-on activities	Total units of analysis
	The Human Body (B)	8	3	11
	Health (B)	2	3	5
	The Environment (PS)	4	5	9
	Conservation of Plants & Animals (B)	5	6	11
	Plants (B)	8	8	16
	Chemical Reactions (PS)	7	3	10
	Light (PS)	5	6	11
	Electric Circuits & Pressure (PS)	10	6	16
	Energy & its Conservation (PS)	6	6	12
	Communication (PS)	7	4	11
	Total	62	50	112

Notes: B denotes that the topic covers biology concepts; PS denotes that the topic covers physical science concepts.

anchoring phrases in the framework, as well as examples of the units of analysis that align to each practice. Since science and engineering practices serve as sense-making contexts/processes that enable students to investigate and make sense of scientific phenomena by building and applying explanatory models and designing solutions to world problems (Schwarz et al, 2017), we believed they were appropriate in helping us map the status of Zambia's science practical work.

Data analysis

Document content analysis, as suggested by Krippendorff (2004), was used to analyze the extent to which the eight science and engineering practices were addressed in the syllabi and textbooks. Content analysis was conducted using line-by-line analysis of inquiry skills statements in the syllabi, and hands-on experiments in the textbooks. The *anchoring phrases* for each description of the practices (see Table 8.2) served as descriptors and guided coders in what to look for during the coding process. Examples of actual experiments and how they were coded for coverage of science and engineering practices are shown in Table 8.3.

The research team consisted of two science education professors. The analysis process consisted of two initial phases of coding and rating of ten randomly selected units of analysis for coders to get familiar with the process. The inter-rater reliability was established between two coders using Cohen's Kappa (Cohen, 1960) at 0.86, which is almost near-perfect agreement. The high Kappa value can be attributed to the presence of the anchoring phrases for each practice that guided the coders, and made the coding process quite consistent. Any disagreements in coding were resolved by having a collective discussion and pinpointing the most closely related anchor phrase. After coding, the coverage of each science and engineering practice was scored for its presence and extent of coverage in percentages. To determine the coverage level, the following threshold percentages were used: If the practice was addressed by 70–100% of the inquiry skills statements or hands-on experiments, it was described as **high coverage**; if it was addressed by 40–69% of the inquiry skills statements or hands-on experiments, the practice was described as **medium coverage**; if it was addressed by 1–39% of inquiry skills statements or hands-on experiments, the practice was described as **low coverage**; and if no inquiry skills statements or hands-on experiments addressed the practice, it was described as **no coverage**.

Results

Coverage of science and engineering practices in syllabi inquiry statements

Table 8.4 shows the percentage coverage trends of the science and engineering practices in the syllabi inquiry skills statements. Specifically, the findings show the following: (a) All inquiry skills statements were framed from a student's perspective – that is, students were to be actively involved in implementing and conducting the practices. For instance, typical inquiry statements would read "students will plan and conduct an investigation, or students will formulate scientific explanations for the observed phenomena". (b) Only one engineering practice (i.e. *designing solutions*) was addressed in the syllabi inquiry skills

Table 8.2 Science and engineering practices with specific anchoring phrases, and units of analysis examples

---

1. Asking questions and defining problems

Science *begins with a question about a phenomenon*, such as "Why is the sky blue?" and seeks to develop theories that can provide explanatory answers to such questions. A basic practice of the scientist is formulating empirically answerable questions about phenomena.

Engineering *begins with a problem, need, or desire that suggests an engineering problem that needs to be solved.* A societal problem such as reducing the nation's dependence on fossil fuels may lead to an engineering problem such as designing more efficient transportation systems. Engineers ask questions to define the engineering problem, determine criteria for a successful solution, and identify constraints.

Example: None of the syllabus inquiry skills statements and textbook activities addressed this practice.

Example: None of the syllabus inquiry skills statements and textbook activities addressed this practice.

2. Developing and using models

Science *uses models and simulations to help develop explanations about natural phenomena.* Models make it possible to go beyond observables and imagine a world not yet seen.

Engineering *uses models and simulations to analyze existing systems or solutions in order to see where flaws might occur or to test possible solutions* to a new problem. Engineers also use models to test proposed systems and to recognize the strengths and limitations of their designs.

Examples: Grade 8 syllabus statement: Students will observe parts of the reproductive system using a model.
   Grade 9 textbook – Activity 43: Students used familiar pictorial models (such as lightbulb, bell, boy skateboarding, and man mowing the lawn) and were asked to use those to identify and describe any energy changes.

Example: None of the syllabus inquiry skills statements and textbook activities addressed this practice.

3. Planning and carrying out investigations

Scientists plan and carry out systematic investigations to *observe, test variables, collect data, and use that data to test existing theories and explanations or to revise and develop new ones.*

Engineers use investigations both *to obtain data essential for identifying or specifying design criteria or parameters, and to test their design* ideas.

Examples: Grade 9 textbook – Activity 12: Required students to investigate how much water their families use. They were directed to study their family's monthly water bills and measurements.

Example: None of the syllabus inquiry skills statements and textbook activities addressed this practice.

4. Analyzing and interpreting data

Scientific investigations produce *data that must be analyzed in order to derive meaning, using a range of tools including tabulation, graphical interpretation, visualization, and statistical analysis to identify the significant features and patterns* in the data.

Engineers analyze data collected in the tests of their designs and investigations; this allows them to *compare different solutions, determine how well each one meets specific design criteria, and identify which design best solves the problem* within the given constraints. Like scientists, engineers require a range of tools to identify the major patterns and interpret the results.

Examples: Grade 8 syllabus statement: Students will predict the birth date given the gestation period.

Grade 9 textbook – Activity 31: Required students to investigate how much water their families consume via monthly consumption bills; then they were asked to find out how much water could be wasted in their neighborhoods due to spillage and leaking pipes.

Example: None of the syllabus inquiry skills statements and textbook activities addressed this practice.

5. Using mathematics and computational thinking

In **science**, mathematics and computation are *fundamental tools for representing physical variables and their relationships.* These ways of thinking allow students to make predictions, test theory, locate patterns/correlations, construct simulations, do statistical analyses, and express/apply quantitative relationships.

In **engineering**, mathematical and computational thinking are integral to design by *allowing engineers to run tests and mathematical models to assess the performance of a design solution before prototyping.*

For example, structural engineers create mathematically based analyses of designs to *calculate whether they can stand up to the expected stresses of use and if they can be completed within acceptable budgets.*

Examples: Grade 8 textbook – Activity 21: Students were provided with a table of various solids (e.g. wood, ice, glass) and various liquids (e.g. milk, water, methylated spirit) with masses, and were asked to predict/evaluate which solid will float or sink in a given liquid.

Example: None of the syllabus inquiry skills statements and textbook activities addressed this practice.

6. Constructing explanations and designing solutions

In **science**, *the construction of scientific theories provides explanations of a phenomenon with available evidence.*

The goal for students is to construct logically coherent explanations that are supported by the available evidence.

In **engineering**, *designing solutions using a systematic approach is used to solve engineering problems based upon scientific knowledge and models of the material world.* Each proposed solution results from *a process of balancing competing criteria of desired functions, technological feasibility, cost, safety, esthetics, and compliance with legal requirements.*

Continued

Table 8.2 (cont.)

Examples: Grade 8 textbook – Activity 22: This introductory activity to the topic of energy asked students to explain how things get hot.

Grade 9 textbook – Activity 1: Required students to explain why people breathe faster when running compared to at rest.

Examples: Grade 8 textbook – Activity 3: Required students to discuss some ways/solutions which can help stop pollution in their communities.

Grade 9 textbook – Activity 17: Required students to brainstorm what people can do to ensure that plants and animals do not become extinct.

7. Engaging in argument from evidence

In **science**, *reasoning and argument with evidence are essential for identifying the strengths and weaknesses of a line of reasoning and for finding the best explanation for a natural phenomenon.* Scientists defend explanations, formulate evidence based on data, and examine ideas with experts and peer understandings.

In **engineering**, *reasoning and argument are essential for finding the best possible solution* to a problem. Engineers collaborate with their peers throughout the design process, with a *critical stage being the selection of the most promising solution, making arguments from evidence to defend their conclusions, and evaluating critically the ideas of others.*

Example: None of the syllabus inquiry skills statements and textbook activities addressed this practice.

Example: None of the syllabus inquiry skills statements and textbook activities addressed this practice.

8. Obtaining, evaluating, and communicating information

In **science**, the major practices are *information seeking* from various sources (e.g. papers, the Internet, symposia, lectures), *evaluation of the scientific validity of the information* acquired, and the *communication of ideas and the results* of inquiry either orally, in writing, with the use of tables, diagrams, graphs, and equations, and by engaging in extended discussions with scientific peers.

Engineers cannot produce new or improved technologies if the advantages of their designs are not communicated clearly and persuasively. Engineers need to be able to *obtain information from various sources (e.g. papers, the internet, symposia, lectures), evaluate the accuracy of the information, and communicate their ideas, orally and in writing, with the use of tables, graphs, drawings, or models, and by engaging in extended discussions with peers.*

Examples: Grade 9 textbook – Activity 16: Requires students to use reference books or the internet to obtain information about plant reproduction.

Grade 9 textbook – Activity 49: Students engaged in a mock interview with a media company representative to obtain information on how a radio works.

Example: None of the syllabus inquiry skills statements and textbook activities addressed this practice.

Note: Italic text represents the *anchoring phrases* for each practice, and served as analysis guides during coding.

Table 8.3 Examples of science practical activities and how they were coded

Experiment 7.3 Investigate the properties of oxygen. (*From Grade 8 textbook, page 116*)	*Experiment 5.5 Show how transpiration works* (*From Grade 9 textbook, page 85*)
Aim: To investigate the properties of oxygen. **You will need:** A sealed test tube filled with oxygen (supplied by your teacher), a wooden splint or match, tongs. **What to do:** 1. Observe and record the colour of the oxygen gas. 2. Carefully smell the oxygen gas by waving your hand over the top of the test tube to sweep some of the gas towards your nose. What does it smell like? 3. Light the wooden splint or match. Blow it out. While it is still glowing, use the tongs to put it inside the test tube. Observe and record what happens. **Results:** 1. Oxygen has no colour and no odour. 2. A glowing splint re-lights in the presence of oxygen. This is the test for oxygen. **Conclusion:** Oxygen is a colourless, odourless gas that supports combustion.	**You will need:** Two potted plants, four clear plastic bags, string or elastic bands **What to do:** 1. Gently remove all the leaves from one of the plants. 2. Cover the soil in each pot with a plastic bag. 3. Use the string to tie another plastic bag around each plant so that the plant is completely covered. 4. Place both plants in the sun and observe what happens after a few hours. **Questions:** 1. What did you see inside each bag? 2. Where did the droplets come from? Explain your answer. 3. What is the process called when a gas turns into liquid? 4. Why were there no drops in the bag around the plant that had no leaves? 5. What does the experiment show?
How coding was done: • What to do section (coded as SP3 for carrying out, not planning the procedure). • Results section coded as no SP6 because explanations were given • Conclusion section coded as no SP4 because analysis/interpretation was given.	**How coding was done:** • What to do section (coded as SP3 for carrying out, not planning the procedure). • Questions 2 and 4 coded as SP6 as they asked students to provide explanations. • Question 5 coded as SP4 as it asked students to provide an analysis or interpretation.

statements, though to a very low extent. (c) In both syllabi, the inquiry skills statements revealed **medium coverage** of *Planning and carrying out investigations*; **low coverage** on several practices in the following descending order: *Obtaining, evaluating, and communicating information; Analyzing and interpreting data; Using mathematics and computational thinking; Developing and using models; Constructing explanations* (for science) and *designing solutions* (for engineering); and **no coverage** of *Asking questions* (for science), and *Engaging in argument from evidence*.

Table 8.4 Percentage coverage of science and engineering practices in syllabi inquiry skills statements

Science and engineering practices	Grade 8 syllabus *Inquiry statements* *(n=56)*	Grade 9 syllabus *Inquiry statements* *(n=62)*	Salient observations
Asking questions (science) and defining problems (engineering)	0	0	• All syllabi inquiry skills statements were written from a student's perspective (e.g. students will plan and conduct experiments). • To the contrary, the hands-on activities in the textbooks (see next subsection) either required students to simply follow the provided experimental procedures without planning, or they were given the analyses and interpretations for the data they collected.
Developing and using models	7	3	
Planning and carrying out investigations	41	49	
Analyzing and interpreting data	18	22	
Using mathematics and computational thinking	7	6	
Constructing explanations (science) and designing solutions (engineering)	5	3	
	2	2	
Engaging in argument from evidence	0	0	
Obtaining, evaluating, and communicating information	30	39	

Notes: (a) Percentages are rounded off to the nearest whole number for easy reading. (b) All practices (except for practice #6) have one row because they only covered the science practices and not the engineering practices. (c) Practice #6 has two rows because it covered both the science practice (top row) and engineering practice (bottom row).

Coverage of science and engineering practices in textbooks

Table 8.5 shows the coverage trends of the science and engineering practices in the textbooks' hands-on experiments, as well as some salient observations worth noting. The findings show the following trends: (a) No engineering practices were covered. (b) In both textbooks, the hands-on activities revealed **high coverage** of *Planning and carrying out investigations*. However, a line-by-line analysis revealed that nearly all hands-on activities only required students to follow the provided experimental procedures, and did not provide opportunities for students to actually plan the procedures. (c) **Low coverage** was found in three practices in the following descending order: *Constructing explanations* (for science) and *designing solutions* (for engineering); *Using mathematics and computational thinking*; and *Developing and using models*. Salient observations showed that some experiments did not require students to construct explanations – instead, they were provided. With respect to *Developing and using models*, students were not asked to *develop* models; instead, they were required to *use* the provided models in the form of diagrams, scenarios, or experimental set-ups. (d) There was **no coverage** of *Asking questions* (for science) and *defining problems* (for engineering); *Engaging in argument from evidence*; and *Obtaining, evaluating, and communicating information*. (e) There was a discrepancy in the coverage of *Analyzing and interpreting data*, as it received medium coverage in Grade 9 and low coverage in Grade 8. A line-by-line analysis showed that most hands-on experiments in Grade 8 provided the analysis and interpretation, while in Grade 9 some experiments did not require students to analyze or interpret data.

Coverage of science and engineering practices across discipline-specific activities

According to Table 8.6, the following findings were revealed. (a) None of the engineering practices were covered. (b) There was a discrepancy in emphasis between the discipline-specific activities. That is, there was **high coverage** of *Constructing explanations* (for science) in life science activities, and **medium–high coverage** of *Planning and carrying out investigations* in physical science activities. (c) The practice, *Analyzing and interpreting data*, had medium coverage in Grade 9 and low coverage in Grade 8. (d) There was **no coverage** of *Asking questions* (for science) and *defining problems* (for engineering); *Engaging in argument from evidence*; and *Obtaining, evaluating, and communicating information*.

Discussion

The primary goal of this content analysis was to determine the nature and extent to which practical work in the Zambian Integrated Science Curriculum integrates the contemporary science and engineering practices. As we discuss

Table 8.5 Percentage coverage of science and engineering practices in textbooks

Science and engineering practices	Grade 8 textbook — Hands-on experiments (n=42)	Grade 9 textbook — Hands-on experiments (n=50)	Salient observations
Asking questions (for science) and defining problems (for engineering)*	0	0	Students were not involved in crafting their own investigative questions or problems.
Developing and using models *	12	9	In both textbooks, students were not engaged in developing the models – instead, they were required to *use* provided experimental set-ups for analysis.
Planning and carrying out investigations *	100	100	In both textbooks, all experiments required students to carry out investigations using the provided procedures. Very few activities required students to *actually plan* (i.e. only activity 30 in Grade 8, and activity 12 in Grade 9).
4. Analyzing and interpreting data *	32	61	In Grade 8, most hands-on experiments (38%) provided the analysis and interpretation. In Grade 9, some experiments did not require students to analyze or interpret data.
Using mathematics and computational thinking	21	17	
Constructing explanations (for science) and designing solutions (for engineering) *	32	39	In both textbooks, some experiments did not require students to make explanations.
Engaging in argument from evidence	0	0	
Obtaining, evaluating, and communicating information	0	4	

Notes: (a) Percentages are rounded off to the nearest whole number for easy reading. (b) Practices with an asterisk (*) had salient observations worth noting.

Table 8.6 Percentage coverage of science and engineering practices across discipline-specific activities

Science and engineering practices	Grade 8 textbook		Grade 9 textbook		Salient observations
	Life science activities (n=8)	Physical science activities (n=34)	Life science activities (n=26)	Physical science activities (n=24)	
Asking questions (for science) and defining problems (for engineering)	0	0	0	0	
Developing and using models	13	6	19	17	
Planning and carrying out investigations *	50	88	31	58	Nearly all experiments simply required students to follow the procedure, and not actually plan.
Analyzing and interpreting data *	13	35	42	46	Most experiments provided the analysis and interpretation.
Using mathematics and computational thinking*	0	26	4	13	Math skills were mostly addressed in physical science activities.
Constructing explanations (for science) and designing solutions (for engineering)	75	35	70	42	
Engaging in argument from evidence	0	0	0	0	
Obtaining, evaluating, and communicating information	0	0	0	0	

Notes: (a) Percentages are rounded off to nearest whole number for easy reading. (b) Practices with an asterisk (*) had salient observations worth noting.

the results, we would like readers to know that at the time our study was conducted, there was no known study that had been conducted on this topic in Zambia. Therefore, our discussion will draw comparisons from some studies that investigated some aspects of science and engineering practices in similar or dissimilar contexts. Our study revealed the following trends:

a   A salient observation between the units of analysis (i.e. syllabi inquiry skills statements and textbook hands-on experiments) revealed a *mismatch in emphasis* in that all inquiry skills statements were written from a student's perspectives (e.g. students will plan and conduct experiments), while the hands-on activities either required students to simply follow the provided experimental procedures without planning, or they were given the analyses and interpretations for the data they collected. Gott and Duggan (2007) assert that if the emphases of science practical tasks are not clearly matched and specified in the curriculum, there is a natural tendency for curriculum developers, textbook writers, and teachers to overlook them. This was the case with our findings in which textbook writers did not frame the hands-on activities so as to emphasize that students be actively involved in implementing the science and engineering practices.

b   In all units of analysis (syllabi inquiry skills statements and textbook hands-on experiments), *nearly all the science practices were addressed*, though to varying extents. To the contrary, *all engineering practices were not covered* except for *Designing solutions practice* that was addressed to a very low extent in the syllabi inquiry skills statements only. These results are similar as well as dissimilar to recent studies. For example, Meyer et al (2012) support our results as they also found that science activities that included engineering practices lagged behind those including science practices. On the other hand, some recent studies (e.g. Chabalengula et al, 2017; Meyer et al, 2012; Moore et al, 2015) found that both science and engineering practices were covered in the materials analyzed, though to varying degrees.

c   In all units of analysis, there was **no coverage** of *Asking questions* (for science) and *defining problems* (for engineering); and *Engaging in argument from evidence*. These findings are similar to recent findings by Chabalengula et al (2017) in which *Asking questions* (for science) and *defining problems* (for engineering), and *Engaging in argument from evidence* were not addressed in widely used K-12 engineering programs in the United States.

d   In all units of analysis, there was **low coverage** of *Constructing explanations* (for science); *Using mathematics and computational thinking*; and *Developing and using models*. Salient observations from a line-by-line analysis revealed that quite a number of experiments did not require students to construct explanations – instead, explanations were provided. With respect to *Developing and using models*, several activities did not require students to **develop** models; instead, they were required to *use* the provided models in the form of diagrams, scenarios, or experimental set-ups.

e   In all units of analysis, there was **medium–high coverage** of *Planning and carrying out investigations* (i.e. high coverage of hands-on activities, and medium coverage of inquiry skills statements). However, a line-by-line analysis of the hands-on activities revealed that nearly all hands-on activities only required students to "carry out/follow" the provided experimental procedures, and did not provide opportunities for them to come up with their own plan and design for the experiments.

f   There were **discrepancies** in the coverage on *Analyzing and interpreting data* and on *Obtaining, evaluating, and communicating information*. With respect to *Analyzing and interpreting data*, the inquiry skills statements in both syllabi and Grade 8 textbook experiments showed low coverage, while there was medium coverage in Grade 9 textbook activities. A salient observation via a line-by-line analysis showed that most hands-on experiments in Grade 8 provided the analysis and interpretation. With regard to *obtaining, evaluating, and communicating information*, the inquiry skills statements strived to cover this practice (though to a low extent), but it was not covered in hands-on experiments in both textbooks.

g   In discipline-specific hands-on experiments (i.e. life science and physical science oriented activities), there was **no coverage** of *Asking questions* (for science) and *defining problems* (for engineering); *Engaging in argument from evidence*; and *Obtaining, evaluating, and communicating information*. Chabalengula et al (2017) also found that across science discipline units (life, physical, and earth sciences units), there was low to no coverage on the aforementioned practices.

h   Life science activities had **high coverage** of *Constructing explanations* (for science), while physical science activities had **medium–high coverage** of *Planning and carrying out investigations*.

Based on our findings, the following aspects merit discussion. First, the lack of engineering practices coverage is a point of concern as it shows that these materials do not provide learners with a sense-making context (in this case an engineering context), in which students can apply their science knowledge and inquiry skills to identify societal problems and design solutions to those problems. This finding is supported by Osborne and Quinn (2017) who stated that many inquiry-based science activities emphasize investigations without any links to real-world problems. Second, not engaging students in *Asking questions* (for science) and *defining problems* (for engineering); *Engaging in argument from evidence*; and *Obtaining, evaluating, and communicating information* in hands-on experiments shows that the Zambian Integrated Science Curriculum materials are overlooking what we consider to be the "foundational" practices, which are critical in initiating and framing the scientific investigations and engineering design problems. We argue that for students to be able to engage in higher-order practices such as constructing plausible explanations or analyzing and interpreting data, they must have engaged in other foundational practices, particularly scientific questioning and defining engineering problems, gathering evidence via carrying out investigations or other scientific texts to obtain data/information to serve as evidence for a line of reasoning for explaining natural phenomena, and using models to develop explanations about natural phenomena. Similarly, without well-defined engineering problems, students may not be able to successfully embark on higher-order engineering practices such as designing appropriate engineering solutions. Other recent studies (e.g. Moore et al, 2015) have also alluded to the idea that general problem-solving skills are prerequisites to solving engineering problems, and students should be able to formulate a design plan as well as identify the need for engineering solutions. Third, a low emphasis on *Analyzing and interpreting data*, particularly in the syllabi inquiry skills statements, may negatively influence how teachers and textbook writers may implement this practice. In fact, the results revealed an undesirable situation in which because the syllabi did not emphasize the practices, a logical trend was observed in which textbook writers did not also put emphasis on the practice. Therefore, curriculum materials such as syllabi which are primary sources of science teaching and learning need to be designed as exemplars for the coverage of scientific and engineering practices, and content that needs to be taught or learned. Fourth, for students to be adept at applying the science and engineering practices as they progress through the education spectra, it is critical that these practices are integrated in all discipline-specific science subjects such as life and physical sciences. Unfortunately, our study revealed discrepancies in coverage emphases between life and physical science activities, with the former emphasizing *Constructing explanations* (for science) and the latter emphasizing *Planning and carrying out investigations* (for science). Further, line-by-line analyses revealed that nearly all hands-on activities only required students to "carry out/follow" the provided experimental procedures, without engaging them in actually coming up with their own plan and design for the experiments. In addition, salient observations revealed that quite a number of science experiments provided explanations, analyses, and interpretations of data. We argue that for students to be meaningfully immersed in the science and engineering practices, they need to be fully engaged in the entire process from the onset, beginning with asking questions or defining engineering problems, to planning their own experimental procedures, analyzing and interpreting data, and constructing explanations.

Implications for the design of science practical work activities

The revealed disparities in the coverage of science and engineering practices raise concerns about the design of science practical work activities in Zambia's Integrated Science Curriculum materials. In light of our study results, we argue that the current design of practical work activities in Zambia is too limited and restricts students in developing conceptual understanding, acquiring authentic inquiry skills, and in using practical work as a learning context to apply knowledge. Currently, the design of practical work activities in Zambia is predominantly "procedural in nature", where students simply follow the provided steps to carry out an experiment, with little to no reference to how the investigated phenomena relate to science content learned or on how they relate to personal/societal applications. Students are never involved in generating questions or defining problems for the investigations, but were occasionally required to construct explanations and analyze and interpret data. This design set-up does not provide opportunities for students to deeply engage with the activities at a conceptual level, nor does it enable students to see links between science concepts being illustrated in the activities, the procedural aspects engaged in, and how the activity is relevant to them. In support of our finding, Abrahams and Millar (2008) concluded that practical work activities tend to predominantly make students manipulate physical objects and equipment, and rarely engages students in cognitive challenges for them to link observations and experiences to conceptual ideas or to develop understanding of scientific inquiry procedures. Similarly, Van Uum et al (2016) recently warned that procedures can be executed without reflection on the generation of scientific knowledge.

As such, we propose that a more multifaceted design matrix to practical activities is required that takes into account the following three aspects: (a) enable students to see an explicit link between forms of science knowledge such as conceptual and inquiry procedural knowledge, and how both impact the status of the scientific claims being generated (further details are provided in the next subsection); (b) a structured learning cycle, in this case we are adopting the 5E learning cycle (an extended description of the 5E cycle is provided in the next section); and (c) Integrating the science and engineering practices, which Schwarz et al (2017) describe as a kind of improved scientific inquiry process that articulates more clearly what successful inquiry looks like when it results in conceptual and procedural understandings. We refer to this multifaceted design approach as the Three-Dimensional Design Matrix (3D-DM) for practical work activities. While some researchers have developed some forms of practical work analysis inventories (e.g. Millar, 2009) or a framework for practical work in science and scientific literacy (Gott & Duggan, 2007), our proposed 3D-Design Matrix is unique as it juxtaposes the science and engineering practices onto the knowledge dimensions and the learning cycle (see Table 8.6). We argue that juxtaposing science and engineering practices onto the three forms of science knowledge and the activity learning cycle is invaluable, as it explicitly shows where and when to integrate the practices during practical work activities at various stages of the science tasks. Therefore, we believe the 3D-DM represents a versatile matrix as it guards against merely following step-by-step procedures, but would promote students' conceptual, procedural, and strategic thinking, as well as providing a clearer idea of what practices and knowledge to watch for if evaluating students' performance.

Role of forms of science knowledge in practical work activities: The Program for International Student Assessment (PISA) science framework (OECD, 2019) describes three science competencies that have a direct bearing on science practical work activities. These are: (a) Explain phenomena scientifically, which requires learners to recognize, offer, and evaluate explanations for a range of natural and technological phenomena. (b) Evaluate and design scientific inquiry, which requires students to describe and appraise scientific investigations and propose ways of addressing questions scientifically. (c) Interpret data and evidence scientifically, in which learners engage in analyzing and evaluating data, claims, and arguments in a variety of representations and draw appropriate scientific conclusions. For students to be able to engage meaningfully with all the aforementioned science competencies, they need to utilize three forms of science knowledge: content, procedural, and epistemic knowledge bases. For example, the competency to explain scientific phenomena does not only require a learner to draw upon the content knowledge he/she has acquired, but also on the understanding of the standard procedures of how that knowledge has been derived (procedural knowledge), as well as on the roles and functions of those procedures in justifying the knowledge produced (epistemic knowledge). Similarly, when designing/evaluating scientific inquiry, it is not sufficient to only understand the procedures applied to obtain data (procedural knowledge); it is also necessary to be able to judge whether the procedures are appropriate and the ensuing claims are justified (epistemic knowledge).

Relationship between forms of science knowledge and science and engineering practices

According to OECD (2019), content knowledge refers to the knowledge of the facts, concepts, ideas, and theories about the natural world that science has established. An example would be students' understanding of how plants synthesize complex molecules using light and carbon dioxide. The competency to explain scientific phenomena is thus dependent on knowledge of the facts or theories about the natural environment. With respect to science practical activities, content knowledge is integral in facilitating some science and engineering practices such as generating scientific questions to investigate (for science), defining an engineering problem/need that needs to be solved (for engineering), constructing scientifically correct explanations, designing scientifically informed solutions to world problems (for engineering), and arguing or evaluating information with scientific evidence to provide a line of reasoning. Procedural knowledge refers to the standard procedures that include diverse methods and practices that scientists use to establish scientific knowledge. Glaesser et al (2009) specifically state that procedural knowledge includes knowing how to use knowledge or knowing how to proceed, particularly during science investigations. Examples of procedural knowledge include formulating scientific questions to be investigated or engineering problems to be solved, planning and carrying out scientific investigations and engineering solutions, the control of variables, analyzing and interpreting data, using math and computational thinking, and procedures for obtaining and communicating the information. Epistemic knowledge refers to an understanding of the specific and common practices of scientific inquiry, the process of knowledge building in science, and the status of the scientific claims being generated (Duschl, 2008). Examples of epistemic knowledge include understanding the roles that questions, observations, theories, hypotheses, models, arguments, scientific investigation processes, and peer reviews play in science and in establishing knowledge that can be trusted. In science practical activities, both procedural and epistemic knowledge bases play critical roles that include identifying scientific questions to investigate, ensuring that appropriate inquiry skills are utilized to gather data reliably, judging the appropriateness of the procedures to plan and conduct investigations, ensuring that the data is interpreted with reference to the underlying scientific ideas and the observed data, and whether conclusions from the data are justifiable and communicated appropriately. Collectively, conceptual, procedural, and epistemic knowledge domains are essential to integrate in practical work in order to make it a meaningful learning context.

The 5E learning cycle

One of the rationales for adopting the 5E cycle is that it has recently been found to increase goal orientation (Mupira & Ramnarain, 2018), an aspect that could address some of the shortfalls of practical work cited in the literature – particularly that students were unable to describe what they learned from the practical task or why they undertook the task (e.g. Abrahams, 2017). This cycle has five phases (Bybee et al, 2006) that can be applied at several levels in the design of practical work activities. As an inquiry-oriented approach, the cycle naturally incorporates the science and engineering practices in its phases, and explicitly shows *which*, *when*, and *where* different science and engineering practices can be integrated in the science activity. The *Engagement phase* engages students in the learning task by focusing on an object, question, problem, situation, or event. The activities of this phase make connections to past experiences and expose students' misconceptions. The science and engineering practices on asking a question (for science), defining a problem (for engineering), and designing solutions for a problematic situation are all ways to engage the students and focus them on the instructional task that can either be minds-on or hands-on. The *Exploration phase* is concrete, mostly hands-on in nature, and is meant to afford students opportunities to explore the concept/ideas at hand. The use of tangible materials and concrete experiences is essential. Exploration activities are designed to assist students in generating knowledge, formulating concepts, processes, and skills. The teacher's role in the exploration phase is that of facilitator or coach; they initiate the activity and allow the students time and opportunity to investigate objects, materials, and situations based on each student's own ideas of the phenomena. The *Explanation phase* involves students in the process of constructing explanations for the observed phenomena. In this phase, the teacher directs students' attention to specific aspects of the engagement and exploration experiences in order to aid students to come up with coherent and logical explanations. In the *Elaboration phase*, students have opportunities to further experiences that extend or apply the concepts, processes, or skills. This phase facilitates the transfer of concepts to closely related but new situations or problems. With respect to science and engineering practices, this phase would engage students in information-seeking activities, identifying and executing design solutions to the task, and presenting and defending their approaches to the instructional task. The *Evaluation phase* is the important opportunity for students to use the skills they have acquired and evaluate their understanding. In addition, the students should receive feedback on the adequacy of their explanations. Informal evaluation can occur at the beginning and throughout the 5E cycle. As such, all the eight science and engineering practices are applicable in this phase.

The 3D-Design Matrix for science practical work activities

The 3D-Design Matrix (3D-DM) for science practical work activities provides a framework for describing what the goals of a given practical activity are, as well as targeting the specific science and engineering practices applicable depending on the goals of the activity. The 3D-DM incorporates five inter-dependent phases of the learning cycle (engage, explore, explain, elaborate, and evaluate), which are associated with three forms of science knowledge (conceptual, procedural, and epistemic). Together, the five phases of the learning cycle and the three domains of science knowledge create a matrix of 15 cells, as illustrated in Table 8.7. At this juncture, some may be wondering about the value of this matrix to science educators, curriculum designers, and students! As already highlighted earlier in the chapter, science education literature shows that practical work is unproductive as currently practiced, and that many students cannot articulate what they learned from the practical task. In support of the literature, our current study findings revealed that practical work activities do not actively engage students in formulating their own experimental procedures and rarely engage them in crafting their own investigation questions, constructing explanations, or in analyzing and interpreting data. It is against this backdrop that we believe the 3D-DM will help address and clarify some of the concerns raised in the literature. Specifically, we believe this design matrix will: (a) Explicitly alert stakeholders such as teachers, curriculum designers, and students that practical work is a great learning context to not only enhance procedural understandings, but also conceptual and epistemic understandings. This would ensure that the activities are designed in such a way that explicitly makes links between the observations and the scientific ideas that they were designed to illuminate, and hopefully enable students to see why they did the activity.

Table 8.7 3D-Design Matrix for science practical work activities

3D-Design Matrix for science practical work activities		Domains of science knowledge		
		Conceptual Science activity informs students about the current understandings of scientific phenomena.	*Procedural* Science activity creates opportunities for students to develop inquiry procedures during investigations.	*Epistemic* Science activity combines conceptual and procedural knowledge with scientific reasoning and critical thinking to develop scientific knowledge.
5E learning cycle	**Engage** Deeply involves students in the learning task by focusing on the question, problem, situation, or event	**Engage-Conceptual:** Students are conceptually focused on crafting investigative questions, defining engineering problems, analyzing scenarios/situations. Applicable science and engineering practices: SP/EP 1, SP/EP 4	**Engage-Procedural:** Students are procedurally focused on crafting investigative questions, defining engineering problems, how to develop and use models, planning investigation procedures, how to conduct investigations, analyze/interpret data. Applicable science and engineering practices: SP/EP 1, SP/EP 2, SP/EP 3, SP/EP 4, SP/EP 5, SP/EP 8	**Engage-Epistemic:** Students are deeply involved in the learning task by focusing on the question, problem, situation, or event from both conceptual and procedural perspectives. Applicable science and engineering practices: SP/EP 1, SP/EP 2, SP/EP 3, SP/EP 4, SP/EP 5, SP/EP 8
	Explore Students interact with materials or data to generate knowledge, formulate concepts, processes, and skills	**Explore-Conceptual:** Students explore with materials/data via models, thought experiments, and minds-on activities, variable relationships to generate knowledge, formulate concepts, processes, and skills. Applicable science and engineering practices: SP/EP 2, SP/EP 3, SP/EP 4, SP/EP 5, SP/EP 7, SP/EP 8	**Explore-Procedural:** Students interact with materials/data via hands-on experiments, physical manipulatives, simulated environments, modeling, science texts, to generate knowledge, formulate concepts, processes, and skills. Applicable science and engineering practices: SP/EP 2, SP/EP 3, SP/EP 5, SP/EP 8	**Explore-Epistemic:** Students interact with materials/data to generate knowledge, formulate concepts, processes, and skills from both conceptual and procedural perspectives. Applicable science and engineering practices: SP/EP 2, SP/EP 3, SP/EP 4, SP/EP 5, SP/EP 7, SP/EP 8

Continued

Table 8.7 (cont.)

3D-Design Matrix for science practical work activities	Domains of science knowledge		
	Conceptual Science activity informs students about the current understandings of scientific phenomena.	*Procedural* Science activity creates opportunities for students to develop inquiry procedures during investigations.	*Epistemic* Science activity combines conceptual and procedural knowledge with scientific reasoning and critical thinking to develop scientific knowledge.
Explain Students provide explanations to questions, problems, situations, or events (from engage phase), and provide explanations to concepts, processes, or skills (from explore phase).	**Explain–Conceptual:** Students explain their question/problem/situation/event or concepts, processes, and skills using various tools (models, graphs, simulated environments) to generate knowledge, formulate concepts, interpret data, and explain relationships. Applicable science and engineering practices: SP/EP 1, SP/EP 2, SP/EP 3, SP/EP 4, SP/EP 5, SP/EP 7, SP/EP 8	**Explain–Procedural:** Students explain their procedure on how they will identify and generate questions/problems, develop/use models, analyze/interpret data, or obtain relevant information. Applicable science and engineering practices: SP/EP 1, SP/EP 2, SP/EP 3, SP/EP 4, SP/EP 5, SP/EP 8	**Explain–Epistemic:** Students provide explanations to questions, problems, situations, or events (from engage phase), and provide explanations to concepts, processes, or skills (from explore phase) from both conceptual and procedural perspectives. Applicable science and engineering practices: SP/EP 1, SP/EP 2, SP/EP 3, SP/EP 4, SP/EP 5, SP/EP 7, SP/EP 8
Elaborate Students have opportunities to extend or apply the concepts, processes, or skills.	**Elaborate–Conceptual:** Students apply the concepts, processes, or skills to define problems, develop models, explain relationships, analyze/interpret data, design solutions, and argue with evidence. Applicable science and engineering practices: SP/EP 1, SP/EP 2, SP/EP 4, SP/EP 5, SP/EP 6, SP/EP 7, SP/EP 8	**Elaborate–Procedural:** Students apply the concepts, processes, or skills to explain how they defined problems, designed models, developed models, argued with evidence. Applicable science and engineering practices: SP/EP 1, SP/EP 2, SP/EP 3, SP/EP 7, SP/EP 8	**Elaborate–Epistemic:** Students have opportunities to extend or apply the concepts, processes, or skills from both conceptual and procedural perspectives. Applicable science and engineering practices: SP/EP1, SP/EP 2, SP/EP 3, SP/EP 4, SP/EP 5, SP/EP 6, SP/EP 7, SP/EP 8

Evaluate
Students have an opportunity to use the concepts and skills they have acquired to demonstrate their understanding.

Evaluate-Conceptual:
Students use concepts, processes, or skills they have acquired to demonstrate their understanding of science concepts and phenomena.
Applicable science and engineering practices:
All 8 practices

Evaluate-Procedural:
Students use concepts, processes, or skills they have acquired to demonstrate their understanding and ability to conduct investigations using science and engineering practices in general.
Applicable science and engineering practices:
All 8 practices

Evaluate-Epistemic:
Students have an opportunity to use the concepts and skills they have acquired to demonstrate their understanding from both conceptual and procedural perspectives.
Applicable science and engineering practices: All 8 practices

Notes: SP refers to Science Practice; while EP refers to Engineering Practice

For teachers, we believe the delineated knowledge bases (conceptual, procedural, and epistemic) would help them to create holistic assessments that take into account students' conceptual understanding, as well as how well they conducted/performed the activity. (b) Help teachers and designers to clearly see which science and engineering practices can be applied to each activity depending on the objectives of the activity. This clarity would help make practical work be organized, productive, and easy to assess as there would be well-defined practices for students to engage with. (c) Help clarify the cognitive demand of the activity based on the learning cycle phases. We believe that incorporating the 5E phases would offer students a variety of ways to express their understandings, and hopefully enable them to see the value of science tasks.

Conclusions

Our study revealed disparities in the coverage of science and engineering practices in practical work activities. These disparities raise concerns about the design of practical work activities in the Zambian Integrated Science Curriculum or any other curricula initiatives elsewhere. We argue that scientific and engineering problems in society are multidisciplinary and multifaceted in nature. As such, an integrated science curriculum should by its nature of being "integrated" aim at nurturing young and adult problem-solvers who can conceptualize problems and solutions in various contexts. Additionally, science curriculum materials should be designed in such a way that science and engineering practices are equally integrated and represented in all aspects, including syllabi learning outcomes and science activities if students are to have a wider sense-making context. Our suggested 3D-DM for science practical activities could be one of the approaches that can explicitly ensure that as many practices as possible are integrated at various stages of the science practical tasks. We hope such a design matrix would support and encourage teachers, curriculum designers, and textbook writers to approach science activities from a multifaceted lens, and enable students to have a clearer picture of what is required of them during science activities.

References

Abrahams, I. (2017) Minds-on practical work for effective science learning. In K. S. Taber & B. Akpan (Eds.), *Science education: New directions in mathematics and science education* (pp. 403–413). Rotterdam: Sense Publishers.

Abrahams, I., & Millar, R. (2008). Does practical work work? A study of the effectiveness of practical work as a teaching and learning method in school science. *International Journal of Science Education*, 30(14), 1945–1969.

Abrahams, I., & Reiss, M. J. (2012). Practical work: Its effectiveness in primary and secondary schools in England. *Journal of Research in Science Teaching*, 49(8), 1035–1055.

Akuma, F. V., & Callaghan, R. (2019). Teaching practices linked to the implementation of inquiry-based practical work in certain science classrooms. *Journal of Research in Science Teaching*, 56(1), 64–90.

Allen, M. (2012). An international review of school science practical work. *Eurasia Journal of Mathematics, Science and Technology Education*, 8(1), 1–2.

Bybee, R. W., Taylor, J. A., Gardner, A., Van Scotter, P., Powell, J. C., Westbrook, A., & Landes, N. (2006). *The BSCS 5E instructional model: Origins, effectiveness, and applications*. Retrieved from www.bscs.org.

Chabalengula, V. M. & Mumba, F. (2012). Inquiry-based science education: A scenario on Zambia's high school science curriculum. *Science Education International*, 23(4), 307–327.

Chabalengula, V. M., Bendjemil, S. A., Mumba, F., & Chiu, J. (2017). Nature and extent of science and engineering practices coverage in K-12 engineering curriculum materials. *International Journal of Engineering Education*, 33(1B), 1–13.

Clackson, S. G., & Wright, D. K. (1992). An appraisal of practical work in science education. *School Science Review*, 74(266), 39–42.

Cohen, J. (1960). A coefficient of agreement for nominal scales. *Educational and Psychological Measurement*, 20, 37–46.

Duschl, R. A. (2008). Science education in three-part harmony: Balancing conceptual, epistemic, and social learning goals. *Review of Research in Education*, 32(1), 268–291.

Ferreira, S., & Morais, A. M. (2014). Conceptual demand of practical work in science curricula: A methodological approach. *Research in Science Education*, 44(1), 53–80.

Glaesser, J., Gott, R., Roberts, R., & Cooper, B. (2009). The roles of substantive and procedural understanding in open-ended science investigations: Using fuzzy set qualitative comparative analysis to compare two different tasks. *Research in Science Education*, 39(4), 595–624.

Gott, R., & Duggan, S. (2007). A framework for practical work in science and scientific literacy through argumentation. *Research in Science & Technological Education*, 25, 271–291.

Hewson, M., & Hewson, P. (1983). Effect of instruction using student prior knowledge and conceptual change strategies on science learning. *Journal of Research in Science Teaching*, 20(8), 731–743.

Hodson, D. (1991). Practical work in science: Time for a reappraisal. *Studies in Science Education*, 19, 175–184.

Kidman, G. (2012). Australia at the crossroads: A review of school science practical work. *Eurasia Journal of Mathematics, Science and Technology Education*, 8(1), 35–47. doi:10.12973/eurasia.2012.815a

Krippendorff, K. (2004). *Content analysis: An introduction to its methodology*. Thousand Oaks, CA: Sage Publications.

Lazarowitz, R., & Tamir, P. (1994). *Research on using laboratory instruction in science*. In D. L. Gabel (Ed.), *Handbook of research on science teaching and learning* (pp. 94–128). New York: Macmillan.

Lunetta, V. N., Hofstein, A., & Clough, M. P. (2007) Learning and teaching in the school science laboratory: An analysis of research, theory, and practice. In S. K. Abell & N. G. Lederman (Eds), *Handbook of research on science education* (pp. 393–441). Mahwah, NJ: Lawrence Erlbaum.

Meyer, D. Z., Kedvesh, J., & Kubarek-Sandor, J. (2012). Creating science and engineering practices in the K-12 classroom: An initial survey of the field, American Society of Engineering Education (ASEE) conference.

Millar, R. (2009). *Analyzing practical activities to assess and improve effectiveness: The Practical Activity Analysis Inventory (PAAI)*. York: Centre for Innovation and Research in Science Education, University of York.

Millar, R., Le Maréchal, J.-F., & Tiberghien, A. (1999). 'Mapping' the domain: Varieties of practical work. In J. Leach & A. Paulsen (Eds.), *Practical work in science education: Recent research studies* (pp. 33–59). Roskilde/Dordrecht, The Netherlands: Roskilde University Press/Kluwer.

Ministry of Education, Science, Vocational Training and Early Education [MESVTEE]. (2015). The Zambian Science Curriculum. Lusaka, Zambia: Oxford University Press.

Moore, T. J., Tank, K. M., Glancy, A. W., & Kersten, J. A. (2015). NGSS and the landscape of engineering in K-12 state science standards. *Journal of Research in Science Teaching*, 52(3), 296–318.

Mulopo, M. M., & Fowler, H. S. (1987). Effects of traditional and discovery instructional approaches on learning outcomes for learners of different intellectual development: A study of chemistry students in Zambia. *Journal of Research in Science Teaching*, 24(3), 217–227.

Mupira, P., & Ramnarain, U. (2018). The effect of inquiry-based learning on the achievement goal-orientation of grade 10 physical sciences learners at township schools in South Africa. *Journal of Research in Science Teaching*, 55, 1–16. doi:10.1002/tea

NRC (National Research Council). (1996). *National Science Education Standards in the USA*. Washington, DC: The National Academies Press.

NRC (National Research Council). (2012). *A framework for K-12 science education: Practices, crosscutting concepts, and core ideas*. Washington, DC: The National Academies Press.

OECD (Organization for Economic Cooperation and Development) (2019). *Programme in International Student Assessment (PISA) 2018 assessment and analytical framework: PISA 2018 science framework*. OECD Publishing. doi:10.1787/f30da688-en

Osborne, J., & Quinn, H. (2017). The framework, the NGSS, and the practices of science. In C. Schwarz, C. Passmore, & B. Reiser (Eds), *Helping students make sense of the world using next generation science and engineering practices*. (Chapter 2, pp. 23–32). Arlington, TX: NSTA Press.

Schwarz, C., Passmore, C., & Reiser, B. (2017). Moving beyond "knowing about" science to making sense of the world. In C. Schwarz, C. Passmore, & B. Reiser (Eds.), *Helping students make sense of the world using next generation science and engineering practices* (Chapter 1, pp. 3–22). Arlington, TX: NSTA Press.

Van Uum, M. S., Verhoeff, R. P., & Peeters, M. (2016) Inquiry based science education: Towards a pedagogical framework for primary school teachers. *International Journal of Science Education*, 38(3), 450–469.

Chapter 9

Ways to implement "practical work" in STEM subjects in Tanzania's secondary schools

Ladislaus M. Semali

Introduction

The term "practicals" refers to science-related activities that students engage in to prepare and enact science laboratory experiments beyond school science (Semali, 2013). The pedagogy of teaching practicals envisaged in the iSPACES method uses an experiential approach in contrast with traditional science curricula outlined by the Tanzanian Education Ministry's Institute of Education in which a mixed framework publicizes teaching science with limited lab experience and sometimes without practicals (Maabula, 2012; Nguru, 2010; O-saki, 2007).

The focus of the iSPACES teaching method, as described in this chapter, is about leveraging STEM principles to create practical solutions for overcoming problems associated with poverty, famine, disease, climate change, and the depletion of non-renewable natural resources. This interdisciplinary project emerged from a quest by stakeholders who analyzed and criticized the degree of practicality of science subjects taught in schools by questioning the teaching methods used in classrooms to engage students and to generate participation in developing solutions for overcoming poverty. The project entailed determining if a community's informal or indigenous science regarding food security can increase agricultural yields, preserve grains, fruits and vegetables, and prevent/treat common diseases affecting humans and livestock as constituents of everyday school science and academic curricula.

To implement practical work in science classrooms, a team of experts designed iSPACES to make explicit the role and procedure of practical work in STEM subjects. The ways of doing science with iSPACES rely on *practicals* that are foundational for experiential learning in and out of classrooms. Experiential learning theory (Kolb & Kolb, 2009) refers to learning as the process whereby knowledge is created through the transformation of experience. The assumption is that knowledge will result from the process of grasping and transforming experience (Kolb, 1984).

The essential factor of iSPACES is *practical work*. A good STEM lesson accomplishes the following strategies:

1   Helps students apply math and science through authentic, hands-on learning
2   Includes the use of (or creation of) technology
3   Involves students in using an engineering design process
4   Engages students in working in collaborative teams
5   Appeals equally to girls and boys
6   Reinforces relevant math and science standards
7   Addresses a real-world problem.(Joly, 2012)

Thus, iSPACES employs curricular thinking that offers teachers a way of teaching science that involves interdisciplinary teaching to motivate students to learn STEM subjects and to produce useful products to solve everyday problems (Semali & Mehta, 2012). The central focus of this thinking is to engage students in *practical work* and to overcome students' *fear of science* subjects, and ultimately to curtail the failure rate in national examinations—of physics, chemistry, biology, and mathematics. In Table 9.1, the failure rate is estimated to be high. Parents and teachers wondered what had gone wrong with students' examination results. The governments of Kenya, Tanzania, and Uganda recognized the challenges posed by a lack of practical work. Reports in the national press regarding poor performance in math and science in national examinations demonstrate the frustration parents, teachers, and policymakers experienced.

The statistics in Table 9.1, though dated, clearly indicate educators' concerns in the methodologies for teaching western science in schools without practicals and the weak performance of students in Kenya, Tanzania, and Uganda. Furthermore, following the 2010 examination results, Uganda's Ministry of Education and Sports resolved to equip teachers' colleges with laboratories and equipment as one way to improve the training of science teachers, recognizing its impact on the development of science and technology (Owuor, 2007).

Table 9.1 Secondary school science performance in Kenya, Tanzania, and Uganda in 2010

Subjects	Kenya (% mean score total 2006)[3]	Tanzania (% failures 2004)[1]	Uganda (% failures 2008)[4]
Physics	40.31	45	58.1[2]
Chemistry	24.91	35[6]	66.8
Biology	27.44	43	37.6
Mathematics	19.01	70	52.5 (2000)[5]

Sources:
1 School Inspection Programme for Secondary Schools in Tanzania (Mabula, 2012, p. 237).
2 Secondary School Science Remedial Programme proposals to facilitators doc (Sumra & Katabaro, 2014).
3 Mainstreaming Gender in science and Technology Policies and Programmes in Kenya UNESCO-NCST, 2010.Nairobi, Kenya: Directorate of Quality Assurance Scheme (QUAS), Ministry of Education.
4 Uganda post-primary Education Sector Report. Africa Region Human Development Working Paper Series / Xiaoyen Liang.
5 The achievement of senior II students in Uganda in English language, mathematics, and biology. Summary of 2010 NAPE report.
6 Reversing the Failure Rate Trends in Science and Mathematics for Tanzanian Schools (Semali, Hristova, & Owiny, 2015).

Problem-solving is the core of STEM-subject investigations. Scholars who broadly support the pedagogy of practical work believe that *practicals* are part of science teaching, and therefore collectively help to position science as distinguishable from other subjects in the secondary school curriculum (Kahn, 1990; Nguru, 2010). Teaching of science with real-world problems in mind drives students' curiosity and investigative interests. But interest in science subjects was found to be low. For example, scholars have found that studies conducted in some developing countries by Munro and Elsom (2000) indicate that only 30% of students studied physics and 32% studied chemistry and mathematics in their higher learning in 1994. These percentages decreased to 25% and 26%, respectively, in 2005 as pointed out in the research by Lyons (2005).

A similar trend can be found in France, Germany, and other developed economies where, for instance, students' enrolment numbers have recurrently decreased at different rates. For example, the trend in some countries shows that Norway experienced a decrease at the rate of 40% from 1994 to 2003, Denmark at the rate of 20% from 1994 to 2002, Germany at the rate of 20% from 1994 to 2001 and the Netherlands at the rate of 6% from 1994 to 2001 (OECD, 2006).

Equally, according to a 2005 Euro barometer study on European schools, trends revealed that the reasons as to why youth in schools were not interested in taking science subjects are complex, and that there is evidence that indicates a strong connection between attitudes towards science subjects and the way in which science subjects are taught (Nguru, 2010). Simultaneously, the reasons for the decrease in the number of students taking science subjects in developed economies vary from the reasons for students in developing economies, particularly in the case of scarcity or lack of qualified science teachers and well-equipped laboratories. Further, studies by Lyons (2005) acknowledged that the decline of interest among young learners in science subjects in Tanzania was a result of how science was taught and learned. This weakness is hard to fix in the short term. Remedies require financial commitment to science education and training of quality teachers.

Similarly, in its policy paper on establishing effective K-12 STEM science education programs, the National Research Council reported that students in high-performing STEM programs had opportunities to learn science, mathematics, and engineering by addressing problems that have real-world applications (Board on Science Education, 2011).

The overarching goal of the discussion in this chapter is to explain the methods for teaching STEM subjects as an integrated curriculum that values the cultural heritage that students bring to science classrooms. Cultural knowledge is practical work. It is rooted in the students' cultural traditions and social environment. Curriculum planners believe that taking integration of practicals seriously in planning physics, chemistry, and biology (PCB) subjects has long-term benefits for students (Beane, 1997). Among these benefits include the opportunity to reduce persistent conflicts between the world of everyday life and the world of school science that minority students and remote area (non-urban) students experience every day in schools (Semali, 2013; Semali, Hristova, & Owiny, 2015).

If we are to advance STEM integration and lift the profile of all of its disciplines, we need to focus on both core content knowledge and interdisciplinary processes. Nations that enjoy high international testing outcomes as well as strong STEM agendas have well-developed curricula that concentrate on 21st-century skills including inquiry processes, problem-solving, critical thinking, creativity, and innovation as well as a strong focus on disciplinary knowledge (Bryan, Moore, Johnson, & Roehrig, 2015; Marginson, Tytler, Freeman, & Roberts, 2013). The need to nurture generic skills, practical work, in-depth conceptual understandings, and their interdisciplinary connections is paramount (Semali & Mehta, 2012). Also explored in this chapter are the capabilities, or lack thereof, that allow teachers to transform the academic environment into a culture that values science as part of students' lives, and consequently, the opportunities for sustainable careers in science. For this reason, this chapter will (a) discuss the value and rationale for practicals in STEM subjects, (b) discuss iSPACES as a teaching method for teaching STEM, and (c) explain the value of "integration" and provide concluding remarks that hone in on the value of *practical work* in secondary schools.

Rationale for teaching practicals in STEM subjects

Practicals in science education characterize the experimental nature of science and the demand for laboratory requirements (Semali, 2013). Generally, teaching STEM subjects with practical work has emerged as a daunting task for pre-service teachers as well as veteran teachers because it is unclear what role practical work should assume. Practicals pose a formidable challenge for several reasons. First, it is difficult to pull together all elements intended to restructure, retrofit, and complement *rather than replace* the current national curriculum in physics, chemistry, and biology (PCB). The first step, however, is to address the common misconception that practicals are too expensive because they require lab space, and costly chemicals. Equally, practicals require many hours of teachers' uncompensated preparation of lab tasks (Semali & Mehta, 2012). Second, practicals are a challenge because they require collecting data that include the results of experiments as well as the lab experience itself, all of which take considerable time to accomplish. Such requirements imply the need to take time to collect data, analyse the data, and write up the report. Third, practicals require the school system's support. Teaching with practicals costs money and can demand considerable investment in laboratory equipment and supplies (Semali & Mehta, 2012).

To appreciate the complexity of teaching with practicals, a physics teacher develops teaching units with basic physics concepts and skills—covering most of the traditional content areas in physics education (e.g., kinematics, mechanics, energy, electricity and magnetism, optics, sound and matter) in which the subject matter deals with personal, social, or scientific contexts. Students and teachers could also add another layer of practical activity, namely producing something useful to solve a problem in daily life (e.g., design a tool or gadget to produce electricity to charge mobile phones in rural areas where there is no grid-electricity).

Given the complexity and cost of implementing practical work in STEM classrooms, governments, educators, policymakers, and stakeholders tend to shun the fully fledged tasks that STEM subjects demand in terms of establishing and running magnet schools or science boot camps. Even though the Ministry of Education mandates teaching physics, chemistry, and biology with practicals, a survey of secondary schools in Tanzania teaching revealed science teaching without practicals (Nguru, 2010). By comparison with other subjects, teaching science subjects is intimidating to pre-service science teachers because science practicals take large amounts of time to reflect and prepare classes in physics, chemistry, or biology. See for example Figure 9.1. For this reason, a teacher must be knowledgeable and able to integrate science concepts with practical applications.

Theoretically, Dewey (1938), Lewin (1936), and Piaget (1957) provided the experiential foundation for iSPACES. Dewey's philosophical pragmatism of experiential education, Lewin's social psychology—based on the vision of "life space" and operating from group dynamics, and Piaget's cognitive-developmental genetic epistemology for understanding and communicating with children, particularly in formal education (e.g., discovery learning), collectively form a unique perspective on learning and child development (Kolb, 1984). These scholars—Dewey, Levin, and Piaget—support the practical dimension of practicals as a necessary condition for inquiry, discovery, and innovation (Van Driel, Beijaard, & Verloop, 2001). Therefore, the quest for new knowledge and discovery, in and of itself, implies the need to search, innovate, and to examine alternatives. All these attempts of trial and error require practical work to experiment.

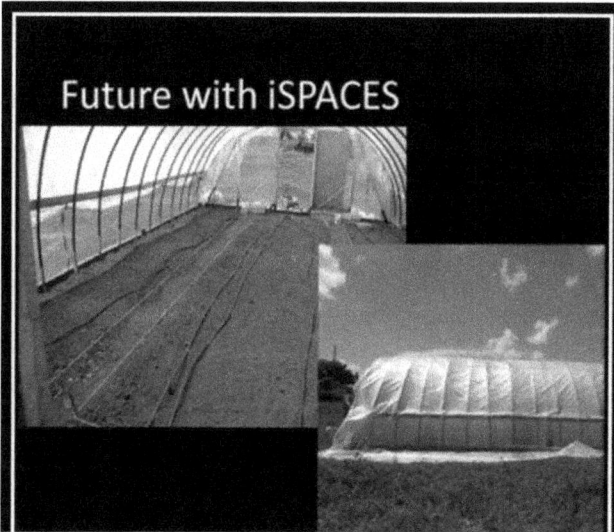

Figure 9.1 Application of STEM skills to build a greenhouse

iSPACES

iSPACES is a way of teaching science that offers an interdisciplinary approach to teaching science as practicals to generate a "to-do" attitude among students (see Semali, 2013, pp. 39–41; Semali, 2017). As an instructional method, iSPACES entails a variety of theoretical, disciplinary, and epistemological perspectives represented in the term iSPACES, to illustrate the interdisciplinary and transdisciplinary nature of learning science purposefully, which includes engineering, education, and the sciences. (See Figure 9.2.)

The iSPACES teaching method is one among many approaches that aim to transform mainstream curriculum by synthesizing appropriate research into pedagogical principles and science content that is readily usable by science teachers. Within this approach, teachers focus on problem-solving to enable graduates of secondary school science to learn and master science concepts, leverage indigenous knowledge, and share, authenticate, enhance, and analyze the knowledge of everyday life. This method encourages students to *think, plan,* and *act* in ways that can help to solve problems or alleviate adverse local conditions (e.g., poverty, lack of clean or safe water, food preservation, etc.). These ways also stress the teaching of science as a powerful tool and useful knowledge that can be used in everyday life (see Table 9.2). But as shown in Figure 9.2,

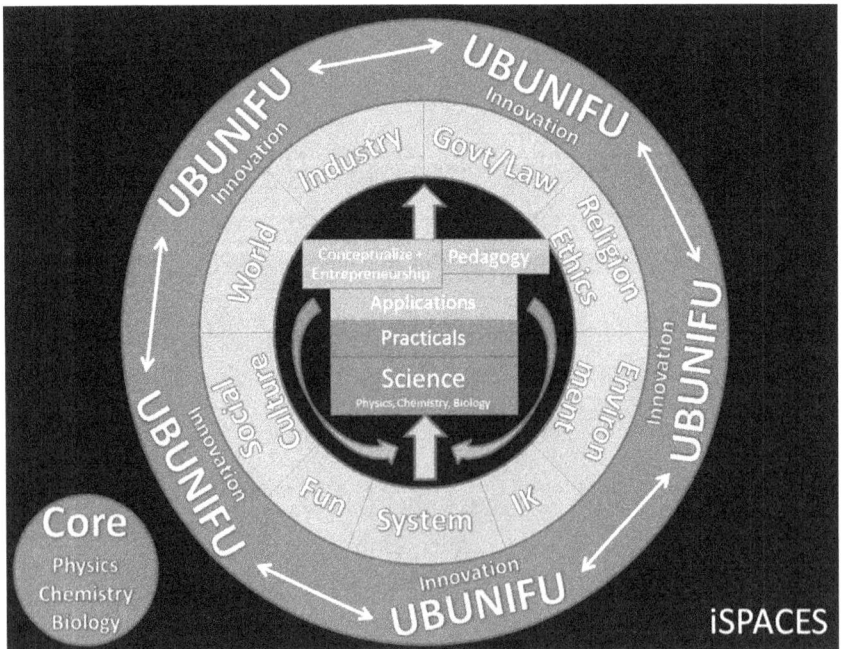

Figure 9.2 iSPACES instructional model for teaching secondary school science (Adapted from Semali & Mehta, 2012, p. 237)

Table 9.2 Conceptualization of the iSPACES teaching method

Concept	Illustration
Science	Students study the concepts/theory of science. (e.g., concept of heat transfer between different materials)
Practicals	Students conduct practicals to reinforce the concepts (Currently, some schools with less resources conduct "alternative to practical" with students memorizing experiments!) (e.g., practical of bi-metallic strip bending with the heat)
Application	Teachers engage students in a discussion on the application of the concept in everyday life. (e.g. heating/cooling food materials)
Conceptualization, Design, and Prototyping	Students conceptualize, design and prototype a simple "machine" based on the science concept. (e.g., designing and building a simple machine to make ice cream)
Entrepreneurship	Students learn how they can employ their design to engage in entrepreneurial activities. (e.g., How do you start a small ice-cream shop?)
Systems	Students learn how the application, design, and entrepreneurial venture relate to the larger system—industry, government and law, religion and ethics, the environment, indigenous knowledge, fun and entertainment, society and culture, and the world at large. This will be delivered through course modules and reinforced through activities related to the iSPACES model. (e.g. how does the ice-cream machine and small business relate to various aspects of the larger "system"?)

Source: Semali and Mehta (2012, p. 237)

Ubunifu must overarch every aspect of the lesson, meaning that the goal of all practicals in science must lead to innovation and discovery.

Even though the study of STEM is widespread, educators' realities include contending with evolving classroom practices to meet 21st-century needs (Sanders & Binderup, 2000). We have indicated elsewhere (see Semali & Mehta, 2012) that the current teaching of STEM education in Tanzania, for example, leaves many students, after four to six years of secondary schooling, without sufficient skills or aspirations for pursuing employment in science-related fields in industry or manufacturing.

The notion of STEM is gradually expanding to include specific disciplines that cover six parts: physics and engineering, chemistry, mathematics, computer science, environmental science, and geosciences. Research shows there is a variety of methods used in these STEM subjects that hone in practicals. Some of the examples include problem-solving strategies of individuals with various levels of physics and expertise (Jonassen, 2000); teaching with cartoons (Keogh & Naylor, 1999); the practical application of what students learn in the STEM classroom (Joly, 2012); transformed curricula that uses hands-on approaches

(Darling-Hammond & Richardson, 2009); expanded problem sets encompassing wonderment (Gilbert, 2013); and laboratory exercises (Bates, 1978).

However, Roth (1991) distinguished two types of practical work that may take another configuration, which could be classified into (a) whole-class practical activities and (b) independent practical activities, where students take on school assignments individually as homework. Whole-class practical activities mainly involve teacher demonstrations of phenomena and objects. The implementation of practical work and the form that whole-class practicals adopt is context-specific. The context may involve a variety of aspects and scenarios. These contexts may include system-wide contexts—e.g. national or local educational policy, status or rank of schools in society, funding, the school system, school governance, and resources, including the science curriculum itself, science equipment, and supplies (Roth et al., 2006).

Equally as important are the individual or personal contexts involving an adequate number of science teachers, parents' education and attitudes towards the hard sciences. Contexts may include science teachers' levels of education, the self-efficacy beliefs of teachers, attitudes towards STEM subjects, professional development of staff, and students' background, life-worlds, aspirations, cultural and social beliefs about science and so on (NSTA, 1992). The ramifications outlined by Roth and colleagues goes to show the daunting complexities and creativity involved in conceptualizing practical work in science classrooms.

Practical work, learning and doing science with the iSPACES method of teaching

Dewey's (1938) thesis on education and experience emphasized experience, experiment, purposeful learning, freedom, and other well-known concepts of progressive education. Dewey believed that sound educational experience involves, above all, continuity and interaction between the learner and the retained knowledge. The goal of the iSPACES method of teaching is to focus on the education of teachers offered at Tanzanian teachers' colleges or in private universities' baccalaureate degrees. The assumption is that if teachers are fired up and engage in practicals, students will have the opportunity to come to love science subjects and will perceive the experiments they conduct as adding value and perhaps as likely to expand what they already know—namely, their prior knowledge.

Experiential learning and the teaching-research pedagogy of critical exploration represent the foundational pillars of the iSPACES method of teaching practicals (Cavicchi, Chiu, & McDonnell, 2009; Duckworth, 2005). To promote such a "to-do" attitude in science classrooms is important and significant for Tanzania, a nation that championed *Education for Self-Reliance* in the early 1970s (Nyerere, 1968). The ways of teaching science framed by the iSPACES method emphasize culturally responsive pathways for learning science, ways which (1) involve teaching practical skills to develop scientific expertise, (2) employ a pedagogy that hinges upon participatory instruction, learning techniques, and innovation, and (3) promote an entrepreneurial spirit and reward critical explorations that focus on solutions to local problems.

To succeed, the teaching of practicals must embrace integration, which is the core theme of the iSPACES teaching method. The traditional science curricula currently promoted by the Tanzania Institute of Education, as observed by Semali and Mehta (2012), involve strict regulation of curricular materials and have established a discipline that disregards the capacities and interests of students' indigenous knowledge, including ignoring conditions and the environment in which students live. This observation is significant to African students who struggle to link what they learn at school in science classrooms with its relationship to what occurs at home in the villages where they live.

As a result of years of devaluing indigenous knowledge and neglect of local examples, science teachers have not managed to address students' cognitive dissonance resulting from the conflict between home science and school science (Semali, Hristova, & Owiny, 2015). For example, locally produced textbooks on African history of science that link what people know and do in their immediate surroundings with current school scientific frameworks for environmental conservation or natural disaster management are difficult to obtain (see Table 9.3).

Practical work in iSPACES

To implement a pedagogy that adopts the iSPACES method of teaching science that is attentive to culturally responsive pedagogy and sensitive to students' *funds of knowledge* and *prior knowledge* (Moll, Amanti, Neff, & Gonzalez, 2005), a

Table 9.3 Example of current lesson objectives and assessment

Main topic	Specific objectives	Assessment
Static Electricity –Concept of Static Electricity	The students should be able to:	
	Explain the concept of static electricity	Is the student able to explain the concept of static electricity?
	Explain the origin of charges	Is the student able to explain the origin of charges?
	Identify the two types of charges	Is the student able to identify two types of charges?
	State the fundamental law of static electricity	Is the student able to state the fundamental law of static electricity?
	Charge bodies using different methods	Is the student able to charge bodies using different methods?

diverse range of stakeholders must be involved. These stakeholders include curriculum planners, classroom teachers, administrators, and the directors of the Institute of Education Planning in the Ministry of Education. Their task is to establish, throughout the development process of PCB lessons, a collaborative effort that includes inputs by local experts, cultural workers, community elders, and practitioners (Semali, 2013, p. 7). Central to the iSPACES thinking is a methodology that allows free exchange of information and perspectives. The classroom teacher understands what knowledge is held locally, valued, and by whom it is held or valued.

Further, to develop the iSPACES method of teaching science, it was necessary to involve cultural experts in the local community and their contributions of knowledge that provide practical solutions to problems. The adjusted thinking resulting from the critical exploration and free exchange of local information was therefore important in creating the nexus and confluence of school science and local knowledge. This approach to curriculum development and teaching of STEM subjects also ensures that teaching strategies undergo experimentation, constant evaluation, challenge, redesign, and perhaps even abandonment when strategies are not applicable.

Shared ways of knowing and familiarity with subject matter helps teachers to use their expertise and authority on behalf of students to make choices that respect each student's sense-making capacities and ultimately nurture each student's interests and development as an individual, rather than striving to maintain fidelity to a programmed national curriculum from the Ministry of Education that is unpractical.

Therefore, the overall assumption undergirding the iSPACES teaching method is that as iSPACES becomes widespread and familiar to many science teachers, the proposed instructional method will continue to need experimentation and fine-tuning. Also, it will need to be situated in the local context (that is, surrounded with cases, models, metaphors, and practical solutions to problems that affect people) where the teaching and learning take place. The pedagogy of iSPACES intent is to demonstrate that (1) indigenous knowledge in science lessons, activities, and class projects adds depth and meaning to difficult concepts and builds communication and respect with the local communities; (2) teaching science in conjunction with local traditional knowledge engenders a *sense of place*, and renders science less foreign to students; and (3) learning to participate in science provides a path for graduates' careers that lead to solving common problems.

Integration of practicals in STEM classrooms

The rationale that supports integrated curricula is based on the belief that the separation of school from the "real world" seems to increase when life experiences, prior knowledge, cultural knowledge, intuition, and organizational frames of meaning remain marginalized or extracurricular (Beane, 1997; Bryan,

Moore, Johnson, & Roehrig, 2015). In general, the integration of practicals deals with the extent to which teachers use local examples from the community, data and information from a variety of cultural groups to illustrate key concepts, principles, generalizations, and theories in their academic subjects or disciplines (Emery, 2000).

Curriculum inquiry in the context of integration represents an intuitive process. The discussion is not about a single issue, a factoid or mathematical formula to be memorized. Therefore, the process of curriculum inquiry is not limited to the "home" or the "school" spheres. It is ongoing, and the home and school are laboratories for experimentations throughout the inquiry process. Sometimes including aspects of heritage knowledge can provide this kind of continuum between all spheres of knowledge production and ultimately reduces, in some ways, the gap between home and school (Semali, 2018).

The interest and impetus for integration of African local knowledge, heritage knowledge, local technologies, and local history of science in STEM classrooms arises from a variety of research and educational spheres (Emery, 2000; Metallic & Seiler, 2009; Moll, Amanti, Neff, & Gonzalez, 2005) (see Table 9.4). First, local knowledge draws from indigenous discourses of knowledge production and community development as a way of telling the African narrative about knowledge, education, and development in diverse African cultural voices that represent history, peoples, and their heritage (Odora-Hoppers, 2002). For our purposes here, we draw from the interdisciplinary and the transdisciplinary levels of integration.

Second, African innovation (referred to in Tanzania as "Ubunifu") is a community's informal science that is part of indigenous populations' ways of knowing, thinking, being, and acting, central to a constant renewal process of trial and error engendered in creativity and intuition to solve problems (Nakata, 2002). Together, innovation, informal science, and community innovations form indigenous epistemology, a worldview or simply a repertoire of the African cultural heritage knowledge "that identifies the African people" in Africa and elsewhere in the world (Odora-Hoppers, 2002, p. 2).

Table 9.4 Increasing levels of integration in STEM

Levels of integration	Features
1. Disciplinary	Concepts and skills are learned separately in each discipline.
2. Multidisciplinary	Concepts and skills are learned separately in each discipline but within a common theme.
3. Interdisciplinary	Closely linked concepts and skills are learned from two or more disciplines with the aim of deepening knowledge and skills.
4. Transdisciplinary	Knowledge and skills learned from two or more disciplines are applied to real-world problems and projects, thus helping to shape the learning experience.

Source: Vasquez, Sneider, & Comer (2013, p. 73)

The iSPACES curriculum method of teaching culturally responsive science in Tanzania emphasizes practical skills to develop scientific expertise. It employs a pedagogy that involves participatory teaching and learning techniques, innovation, and critical exploration with an entrepreneurial focus to motivate students to produce useful products to solve real-life problems (Rae & Wang, 2015).

As shown in Figure 9.1, this framework reflects the principles of science, systems, and entrepreneurship, and focuses on critical exploration and practical skills that aim to solve real problems encountered in life. The aim is to produce products that improve lives and comfort (Semali & Mehta, 2012). Besides the core courses of physics, chemistry, and biology, the curriculum maintains a conceptual framework that includes a holistic systems approach encompassing (a) Core Science, (b) Practicals, (c) Applications, (d) Conceptualizations, (Ubunifu—Design and Prototyping), (e) Entrepreneurship, and (f) Systems. As explained elsewhere, *Ubunifu* is the overarching concept of the curriculum that binds together the iSPACES pedagogical framework (Semali, Hristova, & Owiny, 2015).

Conversely, the proposal to integrate African cultural heritage in STEM subjects is new to many educators trained in African teachers' colleges. While some educators and education reformers enthusiastically embrace the potential pedagogical gains from curriculum integration, such as envisaged in iSPACES, students in many African schools have not developed the mental habits necessary for academic inquiry appropriate for genuine integration. Equally, insistence on rote-learning does not create integration.

Inquiry habits include the ability to imagine and value different perspectives, and then to strengthen, refine, enlarge, authenticate or reshape ideas considering those other perspectives (Beane, 1997). Habits include openness to view ideas combined with a skepticism that demands testing those ideas against previous experience, reading, myth, superstition, or belief. Teaching integration includes *a desire to perceive things holistically and to incorporate specific knowledge into larger frameworks*. These intellectual habits represent the ability to synthesize, analyze, evaluate, and argue—to engage ideas actively and to communicate them. (For a detailed discussion of the rationale for curriculum integration, see Semali, 2000, pp. 29–32.)

Elsewhere, I have described the example where soap was considered an appropriate example of combining science and the production of an item that addresses hygiene. The example of soap was one among many ways a teacher can turn around a traditional curriculum to become practical and expose students to the benefits of linking classroom science to practical uses (Semali, 2013). The assumption is to build a science curriculum that avoids what Tipler (1995) called the "ontological reductionism" concept of science (p. 294). (See also, Birtel, 1995.) The holistic concept of iSPACES shuns the idea of a fragmented world or random jigsaw puzzle pieces, both of which assume separateness of discrete subjects, factoids, mathematical formulas, and equations.

This iSPACES vision of science points to an important distinction from existing programs in Tanzania, and stands in contrast to traditional programs that were initially borrowed from abroad, initiated and conceptualized to increase recruitment and train math and science teachers (Maabula, 2012; Nguru, 2010; O-saki, 2007). STEM education in Tanzania has not solved the endemic problems of poverty, ignorance, and disease which hinder social development and human well-being. In fact, a school subject on "poverty" or life skills is not taught in schools. Notably, in this regard, iSPACES emphasizes the connectedness (as shown in Figure 9.1) between industry, religion environment, local knowledge, and so on. Examples, experiments, and other class assignments must connect with the lives of students and the environment in which they live.

Currently, Tanzanian secondary school students and teachers continue to interact with indigenous and curricular-based natural science knowledges but rarely do teachers make attempts to describe, use and ultimately assign value to the natural science knowledges that are transmitted through instruction in the science classroom and the knowledges that are gained through interactions with local, out-of-school contexts (Semali, 2013).

STEM education has not enabled these three million students to find suitable employment to make a living. Some youth often engage in petty trade but lack an entrepreneurial mindset, focusing on short-term subsistence rather than long-term growth and value-creation. They are stuck in the vicious cycle of poverty and hopelessness. Therefore, there is a realization that the iSPACES approach must extend the learning environment of science beyond the classroom and to design the curriculum in such a way as to engage students' learning beyond rote memorization and, in effect, hone in on "practical skills" which are carefully tailored to stimulate innovation, discovery, and entrepreneurship while validating much of students' local knowledge. In this instance, practical skills refer to the activities and assignments determined to help students live a practical life in real-world situations.

In Table 9.5, we examined the pedagogical method exemplified in the lesson which focused on the topic of "simple machines." We observed that students' learning in this lesson on the wheel and axle is not supported by *problem-solving*. There is a disconnect between the use of the "wheel and axle" in daily life and group discussion about the application of the "wheel and axle" in daily life. The assumption that all students in the class may have had an experience with a wheel and axle is not supported by practice. In practical terms, therefore, students using the iSPACES approach would not stop learning with the lesson in the classroom. Students should be encouraged to experiment with a wheel and axle at home and out of the classroom to discover wheels in all moving parts such as motor vehicles, windmills, pulleys, and so on.

Teachers contemplating using iSPACES should consider integrating in their lessons the history of innovation, African science about historical innovations from the local culture, indigenous medicines, and remedies to illnesses that have been used for millennia (see Table 9.6).

Table 9.5 Example of teaching simple machines

Main topic and subtopics	Specific objectives	Teaching/learning resources	Teaching / learning strategies
Simple Machines	*The student should be able to:*		
Wheel and Axle	Describe the structure of a wheel and axle	• Wheel and axle • Bicycle	• The teacher to display a wheel and axle of a bicycle • Students in groups to discuss the main features of a wheel and axle and how it functions.
	Determine the mechanical advantage, velocity ratio, and efficiency of a wheel and axle	• Heavy load	• The teacher to guide students to determine the mechanical advantage, velocity ratio, and efficiency of a wheel and axle system • Students to determine the mechanical advantage, velocity ratio, and efficiency of the wheel and axle.
	Use the wheel and axle in daily life	• Windlass machine • Bicycle	• The teacher to organize for group discussion on the application of wheel and axle in daily life • Students in groups to cite examples of devices which utilize the principle of the wheel and axle.

Source: Semali (2013, p. 236)

Conclusion

This chapter discussed the rationale to restructure an existing curriculum to enable teachers to rethink the methodologies for teaching with science practicals. The chapter discussed how students from diverse backgrounds—social, cultural, or ethnic—can benefit from the proposed iSPACES method of teaching science that attempts to overcome students' cognitive conflicts between their everyday world and the world of academic science. The iSPACES framework explained how inter-connected spheres of knowledge interact and showed how the principles of science, systems, and entrepreneurship can be designed to combine critical exploration and *practical skills* that aim to solve real problems encountered in life.

Table 9.6 Sources of practical concepts to integrate in academic science fields

Ubunifu descriptions	*Western science (Stem subjects)*
1,500 to 2,000 years ago near Lake Victoria, carbon steel was made in blast furnaces. The temperature achieved in the furnaces, 1,8000°C, was much higher than was managed in Europe until modern times (Van Sertima, 1984, p. 9). Fire was first used 1,400,000 years ago in Chesowanja, near Lake Baringo in Kenya (Van Sertima, 1984, p. 293). An iron-ore mine in Swaziland, the oldest found in the world, was dated as 43,000 years old. The ore specularite was used as a cosmetic and pigment (Zaslavsky, 1984, p. 110).	**Physics** **Physics and Earth Science**
Africans were skilled surgeons. In 1879 in East Africa, a European observed an African doctor carry out a caesarean section successfully, using antiseptic techniques, before this type of operation had been done successfully in Europe (Murfin, 1994).	**Biology (Medical Science)**
Extraction of edible oil from Shea Nut tree (YAO) in Lango, Uganda: Sun-dried Shea Nuts (Yao) seeds are separated from the outer shells, mixed with ashes (ashes prevent burning), and fried. The roasted nuts are cleaned off ashes and pounded into a smooth paste. The paste is mixed with water and put on an open fire in a saucepan or pot, and oils drained as they float to the top, and kept in a cool container (pot).	**Food Science**
Cloth-making technology in Buganda, Uganda and animal skin from other parts of Uganda: The inner bark of the Mutuba tree (ficus natalensis) is harvested during the wet season and then beaten with different types of wooden mallets (sometimes mixing with dyes or wrapped in assorted leaves to soften the fabric) to make its texture soft and fine and give it an even terracotta color.	**Chemistry**
Fermented milk from pastoral communities in Kenya, Tanzania, Uganda: Raw milk poured into a gourd or pot, and gourd transferred to a warm place until the milk has soured and coagulated. Fresh batches of milk may be added each day with or without previous removal of whey, until the gourd or clay pot is full. The fermented milk may be consumed as such or churned to produce butter, consumed or sold (Kerven, 1987).	**Food Science**
Medicinal practice and medicinal plants in Kenya and East Africa: Ferns (treat intestinal worms); Thunbergia alata (treat wounds in mouth and tongue); lannea stuhlmannii (headache and stomach pains); Rhus natalensis (treat influenza, abdominal pain, gonorrhea, and hookworm); Carissa edulis (treat indigestion, abdominal pain in pregnant women, roots' decoction treats malaria).	**Botany (Medical Science)**

Continued

Table 9.6 (cont.)

Ubunifu descriptions	Western science (Stem subjects)
Beekeeping in East Africa: African beekeeper's indigenous knowledge dates back millennia and to date the honey consumed in most households is harvested from traditional hives in tropical forests and processed using traditional methods (Muli, Munguti, & Raina, 2007; Muli, Kilonzo & Sookar, 2018).	Natural Science
Traditional herbs/plants: Over 5,000 herbs, plants and roots have been used routinely by traditional healers in the treatment of diarrhoeal diseases (Anokbonggo, Odoi-Adome, & Oluju, 1990).	Biology
Ghee-making in East Africa (pastoral communities): Butter is heated until it is ready as judged by color (light brown for the ghee residue and straw yellow for the melted butterfat). The molten butterfat is decanted and is then termed ghee.	Food Science

Source: Semali and Mehta (2012, p. 37)

The insights drawn from this discussion indicate, first, that iSPACES can leverage STEM science principles to provide workable solutions to overcome everyday problems associated with poverty, famine, and disease in culturally responsive pedagogy. Perhaps an important example and recognition of valuing cultural heritage knowledge is the art of tracing information, which is the hallmark of science and its foundation in inquiry—and therefore should be taught in schools including higher education.

Equally, there is need to encourage teachers and students to develop interest in cultural heritage, which should be part of formal schooling and a vital source of STEM teaching. In earnest, therefore, the process of integration and innovation exhibits itself in myriad forms in African communities, not exclusive to western science situations or classrooms. As presented in Table 9.6, a short summary of the history of indigenous science in East Africa provides a wealth of knowledge neglected and missing in the current science curriculum. This history captures specific community innovations that span the gamut: from medical practices and medicinal plants to surgery to physics, chemistry and astronomy (Lowe, 1988). However, these practical applications of African science are not valued in the established national curriculum nor are their concepts integrated in the current school curriculum.

These examples (see Table 9.6) illustrate the creativity, discovery, and innovative inclinations of Africans that has existed for millennia. None of this history appears in secondary schools, and none of the subjects set for the Ordinary Level Secondary School examinations contain these materials. Nor are students assessed in such materials in national secondary school examinations.

The role of the teacher in implementing integration of cultural heritage knowledge to rethink and to restructure the secondary science curriculum in Tanzania is critical and cannot be overlooked. Equally, contributions of all stakeholders, including contributions from policymakers, industry, parents, teachers, curriculum planners and professional development experts, are essential to make the integration effort yield results.

Finally, the chapter showed that for the integration process to produce the anticipated outcomes, teachers must recognize that culturally responsive pedagogy facilitates and supports the achievement of all students after identifying the strengths that students bring to school, and nurturing and using those strengths to promote student achievement. Arguably, many challenges lie ahead as teachers and students implement any integration experiment in STEM subjects. Despite looming logistical, academic, and fiscal challenges, teachers and students must commit to try. Delays in embracing cultural heritage knowledge in STEM will continue to erase the cultural heritage knowledge of the environment on the part of farmers or hunters and in school-age students the knowledge of the behavior of the animals being tracked. The divergent interests of policymakers should not prevent them from coming to a consensus between classroom knowledge and home or indigenous knowledge.

In sum, if the teacher is instrumental to the success or failure of implementing any curricular innovation in practice (Mitchener & Anderson, 1989), school and science curriculum reforms must have consonance with the realities of programs for educating science teachers (Shymansky & Kyle, 1992). Any educational reform is likely to fail if it cannot rely on a suitably prepared teaching profession ready to execute educational reforms. For this reason, we commend private universities in Tanzania for exploring this important but neglected area of STEM education that is much in need of attention and for prioritizing it in the design of new baccalaureate teachers' degree programs.

References

Anokbonggo, W. W., Odoi-Adome, R., & Oluju, P. M. (1990). Traditional methods in management of diarrheal diseases in Uganda. *Bulletin of the World Health Organization*, 68(3), 359–363.

Bates, G. R. (1978). The role of the laboratory in secondary school science programs. In M. B. Rowe (Ed.), *What research says to the science teacher* (pp. 55–82). Washington, DC: National Science Teachers Association.

Beane, J. (1997). *Curriculum integration: Designing the core of democratic education*. New York: Teachers College, Columbia University.

Birtel, F. T. (1995). Contributions of Tipler's Omega Point Theory. *Zygon*, 30(2), 315–327.

Board on Science Education. (2011). *A framework for K-12 science education: Practices, crosscutting concepts, and core ideas*. Washington, DC: National Academies Press.

Bryan, L. A., Moore, T. J., Johnson, C. C., & Roehrig, G. H. (2015). Integrated STEM education. In C. Johnson, E. Peters-Burton, & T. Moore (Eds.), *STEM road map: A framework for integrated STEM education* (pp. 23–37). New York: Routledge.

Cavicchi, E., Chiu, S. M., & McDonnell, F. (2009). Introductory paper on critical explorations in teaching art, science, and teacher education. *The New Educator*, 5(3), 189–204.

Darling-Hammond, L., & Richardson, N. (2009). Research review/teacher learning: What matters. *Educational Leadership*, 66(5), 46–53.

Dewey, J. (1938). *Experience and education*. New York: Macmillan.

Duckworth, E. (2005). Critical exploration in the classroom. *The New Educator*, 1(4), 257–272.

Emery, A. R. (2000). *Integrating Indigenous knowledge in project planning and implementation*. Ottawa: Canadian International Development Agency KIVU Nature Inc.

Gilbert, A. (2013). Using the notion of 'wonder' to develop positive conceptions of science with future primary teachers. *Science Education International*, 24(1), 6–32.

Joly, A. (2012). *Real-world STEM problems*. Retrieved January 7, 2019 from www.mid dleweb.com/5003/real-world-stem-problems

Jonassen, D. H. (2000). Toward a design theory of problem solving. *Educational Technology Research and Development*, 48(4), 63–85.

Kahn, M. (1990). Paradigm lost: The importance of practical work in school science from a developing country perspective. *Studies in Science Education*, 18(1), 127–136.

Keogh, B., & Naylor, S. (1999). Concept cartoons, teaching and learning in science: An evaluation. *International Journal of Science Education*, 21(4), 431–446.

Kerven, C. (1987). Some research and development implications for pastoral dairy production in Africa. *ILCA Bulletin*, 26, 29–35.

Kolb, A. Y., & Kolb, D. A. (2009). Experiential learning theory: A dynamic, holistic approach to management learning, education and development. In S. J. Armstrong & C. V. Fukami (Eds.), *The SAGE handbook of management learning, education and development* (pp. 42–68). Thousand Oaks, CA: Sage.

Kolb, D. A. (1984). *Experiential learning: Experience as the source of learning and development (Vol. 1)*. Englewood Cliffs, NJ: Prentice-Hall.

Lewin, K. (1936). *Principles of topological psychology*. New York: McGraw-Hill.

Lowe, J. (1988). 'African science' and its meaning for the secondary school curriculum. *DICE Occasional Papers*, 11, 32–46.

Lyons, T. (2005). Different countries, same science classes: Students' experiences of school science in their own words. *International Journal of Science Education*. doi:10.1080/09500690500339621

Maabula, N. (2012). Promoting science subject-choices for secondary school students in Tanzania: Challenges and opportunities. *Academic Research International*, 3 (3), pp. 234–245.

Marginson, S., Tytler, R., Freeman, B., & Roberts, K. (2013). *STEM: Country comparisons: International comparisons of science, technology, engineering and mathematics (STEM) education*. Final report. Melbourne: Australian Council of Learned Academies.

Metallic, J., & Seiler, G. (2009). Animating Indigenous knowledges in science education. *Canadian Journal of Native Education*, 32(1), 115–128.

Mitchener, C. P., & Anderson, R. D. (1989). Teachers' perspective: Developing and implementing an STS curriculum. *Journal of Research in Science Teaching*, 26(4), 351–369.

Moll, L., Amanti, C., Neff, D., & Gonzalez, N. (2005). *Funds of knowledge for teaching: Using a qualitative approach to connect homes and classrooms*. In N. González, L. Moll, & A. Amanti (Eds.), *Funds of knowledge: Theorizing practices in households, communities, and classrooms* (pp. 29–46). Mahwah, NJ: Erlbaum,.

Murfin, B. (1994). African science, African and African-American scientists and the school science curriculum. *School Science and Mathematics*, 94(2), 96–103.

Muli, E., Munguti, A., & Raina, S. (2007). Quality of honey harvested and processed using traditional methods in rural areas of Kenya. *Acta Veterinaria Brno*, 76(2), 315–320.

Muli, E., Kilonzo, J., Dogley, N., Monthy, G., Kurgat, J., Irungu, J., & Raina, S. (2018). Detection of pesticide residues in selected bee products of honeybees (apis melllifera l.) colonies in a preliminary study from Seychelles archipelago. *Bulletin of Environmental Contamination and Toxicology*, 101(4), 451–457.

Munro, M., & Elsom, D. (2000). *Choosing science at 16: NICEC project report*. Cambridge: NICEC.

Nakata, M. (2002). Indigenous knowledge and the cultural interface: Underlying issues at the intersection of knowledge and information systems. *IFLA Journal*, 28(5–6), 281–291.

Nguru, F. B. (2010). *The analysis of practical work and performance in physics at ordinary level secondary schools in Iringa and Njombe districts (Tanzania)* (Doctoral dissertation, University of Dar es salaam). http://localhost:8080/xmlui/handle/1/51

NSTA (National Science Teachers Association). (1992). *Scope, sequence, and coordination of secondary school science: A project of the National Science Teachers Association (Vol. 3)*. Washington, DC: National Science Teachers Association.

Nyerere, J. K. (1968). *Education for self-reliance* (pp. 167–190). Dar es Salaam: Government Printer.

Odora-Hoppers, C. A. (2002). Indigenous knowledge and the integration of knowledge systems. In C. Odora-Hoppers (Ed.), *Indigenous knowledge and the integration of knowledge systems: Towards a philosophy of articulation* (pp. 2–22). Claremont, Cape Town: New Africa Books.

OECD (Organization for Economic Co-operation and Development)Global Science Forum–ECDGSF. (2006). *Evolution of student interest in science and technology studies policy report*. Paris: OECD.

O-saki, K. M. (2007). Science and mathematics teacher preparation in Tanzania. *Journal of International Co-operation in Education (Hiroshima University)*, 8(1), 111–123.

Owuor, J. (2007). Integrating African Indigenous knowledge in Kenya's formal education system: The potential for sustainable development. *Journal of Contemporary Issues in Education*, 2(2), 21–37.

Piaget, J. (1957). *Construction of reality in the child*. London: Routledge & Kegan Paul.

Rae, D., & Wang, C. L. (2015). Entrepreneurial learning: Past research and future challenges. In *Entrepreneurial learning: New perspectives in research, education and practice* (pp. 25–58). Abingdon, UK: Routledge.

Roth, K. J. (1991). Science education: It's not enough to 'do' or 'relate'. *American Educator*, 13(4), 16–22, 46–48.

Roth, K. J., Druker, S. L., Garnier, H. E., Lemmens, M., Chen, C., Kawanaka, T., Rasmussen, D., Trubacova, S., Warvi, D., Okamoto, Y., Gonzales, P., Stigler, J., & Gallimore, R. (2006). *Highlights from the TIMSS 1999 video study of eighth-grade science teaching (NCES 2006–017)*. U.S. Department of Education, National Center for Education Statistics. Washington, DC: U.S. Government Printing Office.

Sanders, M., & Binderup, K. (2000). *Integrating technology education across the curriculum. A monograph*. Reston, VA: International Technology Education Association.

Semali, L. (2000). *Literacy in multimedia America: Integrating media education across the curriculum*. New York: Falmer Press.

Semali, L., & Mehta, K. (2012). Science education in Tanzania: Challenges and policy responses. *International Journal of Educational Research*, 53, 225–239.

Semali, L. (2013). The iSPACES framework to restructure culturally responsive secondary science curriculum in Tanzania. *Journal of Contemporary Issues in Education, 8* (2), 32–46.

Semali, L. M., Hristova, A., & Owiny, S. A. (2015). Integrating Ubunifu, informal science, and community innovations in science classrooms in East Africa. *Cultural Studies of Science Education*, 10(4), 865–889.

Semali, L. M. (2017). Rethinking the existentialist 'crisis of interest' in school science through culturally responsive African curriculum of STEM science. *Advances in Social Sciences Research Journal*, 4(24).

Semali, L. (2018). Integration of cultural heritage knowledge in STEM and why educators must value and support Indigenous ways of knowing in classrooms. In *Indigenous knowledge in Africa: Ways of being, knowing, acting and reading the world*. Proceedings of the 4th Annual International Conference of the African Association for the Study of Indigenous Knowledge Systems (AASIKS) Conference, 2018 (pp. 17–42). Held at Mwenge Catholic University, Moshi, Tanzania, November 5–7, 2018.

Shymansky, J. A., & Kyle Jr, W. C. (1992). Establishing a research agenda: Critical issues of science curriculum reform. *Journal of Research in Science Teaching*, 29(8), 749–778.

Sumra, S., & Katabaro, J. (2014). Declining quality of education: Suggestions for arresting and reversing the trend. Special Issue, No. 631. *Quality Assurance in Education*, 16(2), 164–180.

Tipler, F. J. (1995). *Critical notice: The physics of immortality: Modern cosmology, God and the resurrection of the dead*. London: Macmillan.

Van Driel, J. H., Beijaard, D., & Verloop, N. (2001). Professional development and reform in science education: The role of teachers' practical knowledge. *Journal of Research in Science Teaching: The Official Journal of the National Association for Research in Science Teaching, 38*(2), 137–158.

Vasquez, J. A., Sneider, C. I., & Comer, M. W. (2013). *STEM lesson essentials, grades 3–8: Integrating science, technology, engineering, and mathematics*. Portsmouth, NH: Heinemann.

Chapter 10

Reflections and future directions in the implementation of practical work in African schools

Umesh Ramnarain

Introduction

The purpose of this book is to highlight the current status of school science practical work in African schools. Substantial literature about practical work in school science has been published during the last two decades, especially on the enactment of practical work in Europe and North America. This period has also seen major school science curriculum reform in African countries, with greater emphasis being placed on practical work. The chapter contributions from South Africa, Namibia, Zimbabwe, Malawi, Kenya, Nigeria, Zambia, Tanzania and Uganda present empirical research on approaches to practical work, contextual factors in the enactment of practical work, use of technology, and teacher preparation and professional development in teaching practical work. This chapter is a commentary on each of the above aspects as elucidated by the chapter authors.

In all nine countries, the increasing influence of globalisation has stimulated the need for more relevant school science curricula. However, in all curricula there is a combination of both global and local influence, which is perhaps encapsulated in the term "glocalisation" – a portmanteau concept that combines "globalisation" and "localisation". Globalisation that is driven by the information technology revolution has heightened the demand for higher-order intellectual skills associated with critical endeavour, creativity, solving abstract problems and innovation that lead to new ideas for improvement. At the same time, there is recognition for local relevance. Integrating science with indigenous knowledge has been viewed as a means by which local relevance in science learning can be reflected. A primary reason advanced for the integration of science with indigenous knowledge systems has been that it acknowledges the wisdom and values that people have acquired over the centuries (Ogunniyi & Hewson, 2008). A further claim for the inclusion of indigenous knowledge in the school curriculum is that it can be used for environmental sustainability in non-western societies (Hewson, 1988; Hewson & Hewson, 1988; Odora-Hoppers, 2002; Ogunniyi, 2011), and hence improve quality of life.

Place of practical work in the science curriculum

In all African countries, school science curricula give priority to three components as described by Childs (2015), namely: the facts and concepts of science (content), the nature and processes of science (conduct or process), and the applications of science in society (context). The role of practical work is recognised in science learning, and is largely associated with the curriculum goal of addressing the nature and processes of science. For example, the authors of the chapter on Nigeria underline the importance that is given to practical work by referring to excerpts from the chemistry curriculum where it is stated that teachers will guide learners, for instance, to "perform experiments to determine the solubility of substances... carry out experiments on the removal of hardness by boiling and addition of washing soda..., prepare the solution of common states" (Nigerian Educational Research and Development Council (NERDC), 2009, pp. 17–19). Similarly, in the Malawian curriculum, it is made explicit that learners should be involved in science processes such as asking questions, designing investigations, conducting investigations, collecting and analysing data, and drawing conclusions (drawing conclusions, relating conclusions to hypothesis) (Government of Malawi, 2013). The South African school Physical Sciences curriculum prescribes practical activities for formal assessment as well as recommended practical activities for informal assessment across grades 10 to 12. The document promulgates that "Practical work must be integrated with theory to strengthen the concepts being taught" (Department of Basic Education, 2011, p. 11). The importance given to practical work is also recognised in Namibia, where in the National Curriculum for Basic Education it is stated that science "contributes to the foundation of a knowledge-based society by empowering learners with the scientific knowledge, skills and attitudes to formulate hypotheses and to investigate, observe, make deductions and understand the physical world in a rational, scientific way" (Namibia MoEAC, 2018, p. 13). In Tanzania, the iSPACES framework configures practical work in the context of real problem solving, fostering innovation and entrepreneurship and contributing to the attainment of sustainable development goals. Accordingly, the focus of iSPACES is to leverage STEM principles to create practical solutions for overcoming problems associated with poverty, famine, disease, climate change, and the depletion of non-renewable natural resources.

Approach to practical work

In aspiring towards the goal of practical work there is a predominance of a "hands-on" approach to practical work, with little or no emphasis being placed on connecting the "hands-on" to the "mind-on". In other words, the development of experimental skills and techniques is often seen as a goal of practical work. According to Woolnough and Allsop (1985), the aim of developing such skills is fundamental in science education as "one cannot be a craftsman unless one can manipulate one's tools" (p. 41). Woolnough and Allsop identify observation, measurement, estimation and manipulation as key skills which can be developed by learners doing practical work. Millar (1991) elaborates upon what he refers to as "practical skills" (p. 51). He divides the skills which he feels can be taught and improved into practical techniques: e.g. measuring temperature to within certain limits, separating by filtration or other "standard" procedures; and inquiry tactics: e.g. repeating measurements, tabulating data and drawing graphs in order to look for patterns, identifying variables to alter, control, etc. According to Millar, by developing these skills, learners will develop their "procedural understanding" of science (in contrast to their conceptual understanding). In order for learners to benefit conceptually from practical work, teachers need to scaffold learners in linking observations and experiences to conceptual science ideas. This appears to be a deficit in science classrooms across Africa.

Within the South African school science landscape, despite the strong curriculum imperative for inquiry-based science education, practical work if it does take place often embodies a recipe or cookbook approach, where learners follow a set of instructions for the execution of procedures such as collecting data by measurement or taking down readings, tabulating collected data, drawing graphs, and arriving at a conclusion (Ramnarain, 2010). In essence, what is reflected in school science curriculum policy has not translated into classroom practice. Similarly, it is reported in Chapter 3 that the Zimbabwean science curriculum endeavours to engage learners in inquiry-based practices, with key objectives such as designing and planning investigations, evaluating methods and techniques, and suggesting possible improvements. The authors allude to a study by Munikwa, Chinamasa and Mukava (2011) where it was established that Zimbabwean teachers employed traditional pedagogical approaches such as demonstrations when doing practical work. It was found that most of the teachers (77%) assigned only one design practical for the whole term, while others neglected it. They observed that this experience of practical work for learners failed to adequately prepare them for summative assessment of practical work in A-level and O-level practical examinations where learners are required to demonstrate investigative, manipulative, critical analysis and evaluation skills. In a case study on the enactment of practical work at two Ugandan schools, the author of Chapter 7 reported that the observed lessons on practical work were largely classified at levels 1 and 2 according to a framework by Baillie and Hazel (2003). This suggests that the investigations were structured with learners not involved in the planning and designing, but mainly confined to data collection, data analysis and arriving at a conclusion. In Nigeria, lessons on practical work are characterised by strong teacher-centredness, with a "cookbook" approach appearing to be the norm. The authors of Chapter 4 on Nigeria highlight for us that teachers take the "centre stage" and have a tendency to "explain away" students' observations and dismiss discrepant students' results to experimental error. There is little or no attempt to engage students on their results or in argumentation of evidence presented.

Assessment of practical work

In some African countries, the importance that is attached to practical work as an indispensable aspect of science learning is underlined in it featuring as a core component of formal assessment. In Kenya, practical skills are assessed independently of content knowledge acquired and this is done at the end of the secondary school course programme in Biology, Chemistry and Physics. Statistics provided in Chapter 6 on Kenya show that learners perform poorly in the paper on practical work in Biology, Chemistry and Physics. The author of this chapter proposes urgent intervention measures be set in place to improve the quality of school science practical work. Summative assessment of practical work also exists in Zimbabwe where in A-level Biology, Chemistry and Physics, learners sit for a two-and-a-half-hour paper that tests inquiry skills. A cause for concern flagged in examination diagnostic reports of the Zimbabwe Schools Examinations Council is that the majority of A-level Biology candidates lose marks in the practical examination due to the difficulty they encounter in the design of experiments and the evaluation of experimental results. These poor results are explained as being due to the inordinate amount of time that is spent during practical lessons on the writing of investigation reports. In Nigeria, the senior school certificate examination administered by the West African Examinations Council (WAEC) is comprised of three papers. One of these papers taken by candidates has three questions that focus on the testing of practical work. Two of the questions test quantitative analysis and qualitative analysis skills, respectively, while the third is on practical tasks suggested in each of the teaching syllabi. The questions are weighted to account for 25% of the total mark for the examination. In the South African curriculum, practical activities are prescribed for formal assessment. In Physical Sciences grades 10 and 11, two prescribed experiments, one in physics and the other in chemistry, are done per year for formal assessment purposes. In the programme of assessment tasks for grade 12 that is a precursor to the end-of-year final exit examination, 45% of all assessment tasks are based on a practical activity. These practical tasks are referred to as "experiments", whose purpose is to "verify or test a known theory" (Department of Basic Education, 2011, p. 145).

Teacher preparedness for practical work

Teacher knowledge and pedagogical orientation appear to be key factors in the form of practical work that unfolds and the extent to which practical work takes place in African schools. With regard to pedagogical orientation, the study that is reported in Chapter 5 shows that Malawian teachers assume a direct didactic pedagogical orientation that actualises in them giving learners questions to investigate, providing step-by-step instructions on how to conduct the procedure, and how to record and present data. The authors remark that teachers hold the belief that the role of practical work is for the acquisition of laboratory skills, verification of theory and to aid memory of science content.

There is a suggestion that such a belief emanated from their teacher education where they experienced practical work as being heavily scripted and based on experiments from a practical manual. In Zambia, a study by Chabalengula and Mumba (2012) reflects a similar picture with teachers having a narrow conception of inquiry that is revealed in them giving more priority to evidence gathering and explaining the evidence, with much less emphasis being placed on justifying the explanations and connecting the explanation to scientific knowledge. The authors suggest that there is a deficit in teacher understanding of inquiry, and this could be addressed in both pre-service and in-service teacher education by means of exposing them to authentic scientific inquiry that would involve them "in identifying problems, formulating hypotheses, designing experiments, gathering and analyzing data, and drawing conclusions about scientific problems or natural phenomena" (p. 323).

In Nigerian schools, a major challenge in the implementation of practical work according to the authors is that teachers are poorly qualified in science teaching with them lacking the requisite knowledge and skills for conducting science practical with learners. Similarly, in South Africa, a constraining factor in the effective implementation of inquiry-based science education has been that teachers do not possess the necessary competency that this pedagogy demands. Although resources have been committed for teacher development, especially for in-service teachers, development programmes have failed to shift teacher practice towards inquiry. Practical work is still deeply entrenched in approaches where learners lack autonomy in practical activities. Teachers appear to lack a clear guideline on how to transform their practice. Studies by the author on empowerment evaluation suggest this may be an approach to be considered in steering teachers' practice, albeit incrementally, in a sustainable and self-determined manner towards a progressive approach to practical work (Ramnarain & Modiba, 2013). Empowerment evaluation is designed to help people help themselves and improve their programmes using a form of self-evaluation and reflection (Fetterman, 2001). Critical friendship characterises the mentoring relationship between the teacher (evaluee) and the coach (evaluator). In a study conducted by the author and a fellow researcher (Ramnarain & Modiba, 2013), an empowerment evaluation approach was applied to enable a Life Sciences teacher to reflect upon and refine his curriculum design principles in promoting the scientific literacy of learners through inquiry-based learning.

The importance that is given to inquiry-based learning in Zimbabwean school science curricula means that teachers need to be adequately prepared in this pedagogy. The authors affirm that this is indeed the case in the country, with universities in their teacher development programmes placing a high premium on supporting future graduates. However, a challenge to this status quo is the financial crisis that the country finds itself in that now limits the resources made available to teacher education institutions, especially in the maintenance of functioning laboratories. A further strain is the exodus of the well qualified and experienced to neighbouring countries for better conditions of service.

A question of physical resources and infrastructure for practical work

A common denominator in studies that have been conducted on the implementation of practical work across schools in Africa has been the issue of the availability of practical resources such as laboratory facilities and apparatus, equipment and consumable supplies. This is an issue that is highlighted by all chapter authors. However, the lack of resources has created an opportunity for teachers to demonstrate their innovativeness and creativity through the design and use of low-cost materials for practical work. In Namibia, the tension between curriculum formulation and implementation has been contributed to largely by a lack of resources to do practical activities, especially in under-resourced rural schools. The authors of Chapter 2 present research findings on a case study of 21 in-service science teachers from mostly rural schools who were undergoing professional development by participating in a Bachelor of Education Honours program. The findings reveal that with the necessary support, teachers can effectively exploit readily accessible low-cost materials for the purposes of practical work. The findings of this study and others (Ramnarain & Mamutse, 2016) provide compelling evidence that the use of improvised physical resources can be a panacea to the challenge of the shortage of conventional physical laboratory resources in African countries. In a case study of practical work at two schools in Uganda, the chapter author reported that the teachers made use of local materials such as plastic cups, pins, rubber bands, strings, potatoes, flowers, tendrils, onions, and other plant stems during practical work. This contributed to learners being engaged in understanding the world around them and in mastering the subject content.

It is also noted that the lack of resources impacts on the pedagogical approach adopted by teachers when doing practical work. For example in Kenyan sub-county schools, teachers resort to demonstrations due to a lack of equipment and chemicals. This situation is compounded by the high enrolment at such schools. The differences in the availability of resources at different schools also exacerbates inequalities that exist in the education system. In Kenya, in contrast to the sub-county schools, the so-called National schools have well-established infrastructure, in terms of well-equipped science laboratories, as well as an adequate supply of teaching and learning resources for practical work. This situation is especially prevalent in South Africa, where the legacy of the Apartheid education system has resulted in schools that are quite different in terms of infrastructure and the provisioning of resources. Township and rural schools that are attended by the majority Black population in general lack properly fitted laboratories and accompanying resources for doing experiments. On the other hand, former model C schools (mainly in suburbs), previously designated for White children but now open to all races, have far superior facilities for practical work. Despite huge financial investment being made in upgrading the facilities for practical work in township and rural schools, these schools still lag behind former model C and elite private schools. The availability of resources is a factor that influences the pedagogical orientation of teachers when doing practical work, and due to resources being related to the location of school, this contributes to inequality in science learning experiences for the population groups. The findings of a study by Ramnarain and Schuster (2014) showed remarkable differences between the pedagogical orientations of teachers at disadvantaged township schools and teachers at more privileged suburban schools. The study found that teachers at township schools assumed a strong "active direct" pedagogical orientation that involved the direct exposition of the science content followed by confirmatory practical work, while teachers at suburban schools exhibited a guided inquiry orientation, and provided opportunity for learners to investigate their own ideas via a guided exploration phase. A similar pattern is noted in Malawi where teachers at more privileged grant-aided schools exhibited stronger guided- and open-inquiry orientations than teachers at less privileged Community Day Secondary Schools and Conventional Schools (Ramnarain, Nampota, & Schuster, 2016).

The role of ICT in practical work

In the African countries that formed the focus of this book, there is recognition of the possible affordances of information computer technologies (ICT) in the doing of practical work. However, economic factors inhibit the wide scale uses of ICT. In South Africa, the value of integrating ICT in learning is expressed through the South African White paper on e-Learning where it is stated that "ICTs, when successfully integrated into teaching and learning, can advance higher-order thinking skills such as comprehension, reasoning, problem solving and creative thinking and enhance employability" (Department of Education, 2004, p. 14). A study there showed that the use of simulations in the science classroom did, to a certain extent, reduce the number of misconceptions previously held by learners, and can be a viable cognitive learning tool in enabling learners to investigate their pre-conceptions and thereby effect conceptual change (Ramnarain & Moosa, 2017). In Chapter 7 on practical work in Uganda, the author contends that the use of ICT can possibly address the issue of shortage of physical resources in schools. In Zimbabwe, the role of ICT in teaching and learning is recognised as facilitating personalised learning and instruction, sharing resources, collaborating, and promoting learning outside the classroom (Munikwa, 2016). A study by Ndibalema (2014) that investigated Tanzanian secondary school teachers' attitudes towards the use of ICT as a pedagogical tool revealed that although teachers had positive attitudes towards using ICT as a tool, they failed to effectively integrate it in their teaching. Although the value of ICT in science teaching and learning is recognised, the lack of infrastructure and teachers not having adequate technological and pedagogical knowledge for integration are two constraining factors.

Conclusion

This chapter has highlighted trends and patterns in the enactment and role of practical work across African countries. It is abundantly clear that practical work is regarded as an intrinsic aspect of science education. In all school science curricula, the importance that is apportioned to practical work is evident in the learning outcomes, aims, standards and goals that are formulated. The form of practical work that is strongly advocated is inquiry-based learning. This approach signals a definite paradigm shift from the traditional teacher-dominated to a learner-centred approach. Here, science is depicted as a human activity, and there is a strong emphasis on engaging the learners in the processes of sciences. However, in practice either teacher demonstration or structured inquiry in the form of a "cookbook" approach are prevalent. Due to these teacher-directed pedagogical approaches, learners do not enjoy the much-documented benefits of inquiry such as improved conceptual understanding, forming an understanding of the nature of science and developing higher-order thinking skills such as evaluation, critical thinking and creativity. Unfavourable contextual factors, such as inadequate resources and large classes, and teachers lacking professional development in the implementation of inquiry, serve to inhibit widespread uptake.

For many schools, the transition towards open or even guided inquiry is a bridge too far. Studies by Ramnarain and Hobden (2015) and Rogan and Grayson (2003) suggest that a stepped approach should be pursued in shifting teachers towards an inquiry-based pedagogy. Rogan and Grayson's Profile of Implementation can be used as a tool to establish the level of implementation in science practical work, taking into account the context and capacity of their school. The Profile of Implementation is designed to "offer a 'map' of the learning area, and to offer a number of possible routes that could be taken to a number of destinations" (Rogan & Grayson, 2003, p. 1181). Within the parameter of practical work, such a destination may be open inquiry that affords the learners with an authentic learning experience. Similarly, a learning progression proposed by Ramnarain and Hobden (2015) can be used by teachers to consciously construct and choreograph an environment that supports and grows learner autonomy in doing practical work.

A pleasing trend across countries is that teachers have invoked the agency to overcome contextual obstacles to practical work by being innovative and resourceful in the use of low-cost and readily accessible materials. This dismisses the myth that practical work cannot be taught in schools without fully fledged laboratories with expensive and sophisticated equipment. It is also suggested that the use of such materials may lead to authentic learning experiences (Poppe, Markic & Eilks, 2010; Yeboah, Abonyi, & Luguterah, 2019).

References

Baillie, C., & Hazel, E. (2003). *Teaching materials laboratory classes*. Liverpool: The UK Centre for Materials Education.

Chabalengula, V. M. & Mumba, F. (2012). Inquiry-based science education: A scenario on Zambia's high school science curriculum. *Science Education International*, 23(4), 307–327.

Childs, P. E. (2015). Curriculum development in science – Past, present and future. *LUMAT*, 3(3), 381–400.

Department of Basic Education (2011). *Curriculum and assessment policy statement: Grades 10–12 physical sciences*. Pretoria: Government Printer.

Department of Education (2004). *White paper on e-Learning*. Pretoria: Government Printer.

Fetterman, D. M. (2001). *Foundations of empowerment evaluation*. Thousand Oaks, CA: Sage.

Government of Malawi. (2013). *Biology curriculum*. Lilongwe, Malawi: Malawi Institute of Education.

Hewson, M. G. (1988). The ecological context of knowledge: Implications for learning science in developing countries. *Journal of Curriculum Studies, 20*, 317–326.

Hewson, P. W., & Hewson, M. G. A'B. (1988). An appropriate conception of teaching science: A view from studies of science learning. *Science Education*, 72(5), 597–614.

Millar, R. (1991). A means to an end: The role of processes in science education. In B. E. Woolnough (Ed.), *Practical science* (pp. 43–52). Buckingham: Open University Press.

Munikwa, S. (2016) *An analysis of Zimbabwean teachers' interpretation of the advanced level physics curriculum: Implications for practice*. Unpublished doctoral thesis, University of South Africa, Pretoria, South Africa.

Munikwa, S., Chinamasa. E., & Mukava, M. (2011). A study of teaching advanced level physics practical and solution approach to practical questions. *Journal of Innovative Research in Education, 1*(1), 36–48.

Namibia Ministry of Education, Arts and Culture (MoEAC) (2018). *The National Curriculum for Basic Education*. Okahandja: NIED.

Ndibalema, P. (2014). Teachers' attitudes towards the use of information communication technology (ICT) as a pedagogical tool in secondary schools in Tanzania: The case of Kondoa District. *International Journal of Education and Research*, 2(2), 1–16.

Nigerian Educational Research and Development Council (NERDC) (2009). *Senior secondary education curriculum: Chemistry for SS1–3*. Yaba, Lagos: NERDC Press.

Odora-Hoppers, C. (2002) (Ed.). *Indigenous knowledge and the integration of knowledge systems*. Claremont: New Africa Books.

Ogunniyi, M. B. (2011). The context of training teachers to implement a socially relevant science education in Africa. *African Journal of Research in Mathematics, Science and Technology Education*, 15(3), 98–121.

Ogunniyi, M. B., & Hewson, M. G. (2008). Effect of an argumentation-based course on teachers' disposition towards a science-indigenous knowledge course. *International Journal of Environmental Science Education*, 3(4), 159–177.

Poppe, N., Markic, S., & Eilks, I. (2010). *Low cost experimental techniques for science education: A guide for science teachers*. Bremen: University of Bremen.

Ramnarain, U. (2010). *Teaching scientific investigations*. Northlands: Macmillan, South Africa.

Ramnarain, U., & Hobden, P. (2015). Shifting South African learners towards greater autonomy in scientific investigations. *Journal of Curriculum Studies*, 47(1), 94–121.

Ramnarain, U., & Mamutse, K. (2016). *The use of improvised resources in inquiry-based teaching in South Africa.* Paper presented at the ESERA conference in Helsinki, Finland.

Ramnarain, U., & Modiba, M. (2013). Critical friendship, collaboration and trust as a basis for self-initiated professional development: A case of science teaching. *International Journal of Science Education*, 35(1), 65–85.

Ramnarain, U., & Moosa, S. (2017). The use of simulations in correcting electricity misconceptions of grade 10 South African physical sciences learners. *International Journal of Innovation in Science and Mathematics Education*, 25(5), 1–20.

Ramnarain, U., & Schuster, D. (2014). The pedagogical orientations of South African physical sciences teachers toward inquiry or direct instructional approaches. *Research in Science Education*, 44(4), 627–650.

Ramnarain, U., Nampota, D., & Schuster, D. (2016). Spectrum of pedagogical orientations of Malawian and South African physical science teachers towards inquiry. *African Journal of Research in Mathematics, Science and Technology Education*, 20(2), 119–130.

Rogan, J., & Grayson, D. (2003). Towards a theory of curriculum implementation with particular reference to science education in developing countries. *International Journal of Science Education*, 25(10), 1171–1204.

Woolnough, B. E., & Allsop, T. (1985). *Practical work in science*. Cambridge: Cambridge University Press.

Yeboah, R., Abonyi, U. K., & Luguterah, A. W. (2019). Making primary school science education more practical through appropriate interactive instructional resources: A case study of Ghana. *Cogent Education*, 6(1), 1–14.

Index

Note: Page locators in *italic* refer to figures and in **bold** refer to tables.

For Product Safety Concerns and Information please contact our EU
representative GPSR@taylorandfrancis.com
Taylor & Francis Verlag GmbH, Kaufingerstraße 24, 80331 München, Germany

www.ingramcontent.com/pod-product-compliance
Lightning Source LLC
Chambersburg PA
CBHW060302220326
41598CB00027B/4201